C. Mattheck

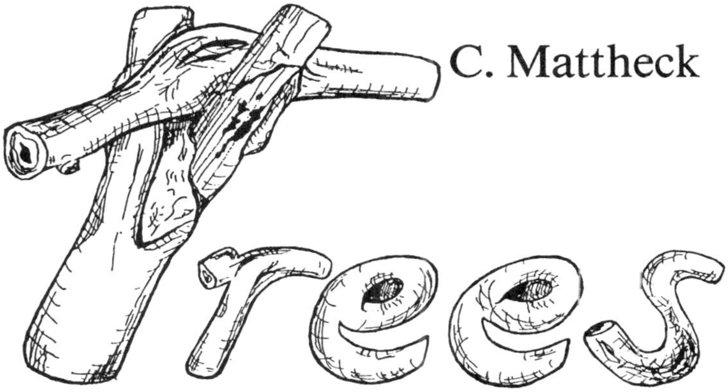

The Mechanical Design

With 107 Figures

Springer-Verlag
Berlin Heidelberg New York
London Paris Tokyo
Hong Kong Barcelona
Budapest

Dr. rer. nat. habil. GERHARD CLAUS MATTHECK
Kernforschungszentrum Karlsruhe
Institut für Materialforschung II
Postfach 3640
7500 Karlsruhe, Germany

ISBN 3-540-54276-0 Springer-Verlag Berlin Heidelberg New York
ISBN 0-387-54276-0 Springer-Verlag New York Berlin Heidelberg

Library of Congress Cataloging-in-Publication Data.

31/3145-543210

To Marita and Willie

ACKNOWLEDGEMENTS

The author would like to thank his enthusiastic collaborators and students who all worked hard at the computer to verify the results, which are presented here in a qualitative way.

The author is also grateful to the management of the Karlsruhe Nuclear Research Centre for the encouraging and stimulating interest in the tree research of his team, for all the continuous promotion, and for the confidence in the success of all activities and efforts.

Some of the rare phenomena and interesting trees presented in this book were detected by attentive scientific partners and friends inside and outside of our country. They provided the author with numerous suggestions and valuable inspiration. The author is grateful to them, as well as to all the friends sharing his interest in this great and vivid working field.

Furthermore, the author would like to thank Mrs. Heidi Knierim who was extremely careful in preparing the manuscript, and who helped considerably with the English translation. Mr. Michael Beller and Dr. Klaus Bethge are gratefully acknowledged for kindly proofreading the English manuscript. Mr. Helmut Zwecker contributed by his careful reproduction of the drawings.

Last, but not least, the author is very grateful to his partners at Springer-Verlag, Heidelberg, whose cooperative support helped to publish this book.

PREFACE

The beauty of a knotty oak tree is different from that of a lovely flower. It is the rough beauty of an old soldier's face showing the traces of wind and sun, of harm and of victory, bearing the scars of bygone battles. It is different from the fragile, delicate beauty of a young girl which is evident to anyone at first sight. The beauty of an old and crippled tree is hidden unless perceived by the alert eye which is able to fancy or rather discern the hard trials of life the tree has experienced.

Contemplating trees in this way is not much different from busying oneself with physiognomics, i.e. with the art of judging character from the features of the human face. Physiognomics is often considered a dubious science, but is practiced every day in human communication by everybody from early childhood to old age. Although we all are able to discern the angrily furrowed brow, the laughing crow's-feet below the eyes, the arrogant harsh lines around the nose, the hard narrow mouth, the gluttonous lip, and the secret eye of the silent observer, we would never admit to rely on such seemingly doubtful methods.

Unlike physiognomics, which is based on comparative experience, the biomechanics of trees is an exact science, which when pursued properly allows one to infer a tree's fate from the individual design it exhibits. In other words, looking at its external shape or the tree rings in a saw cut, one can in most cases trace back the marks and stages of the tree's mechanical fate, i.e. its load history and survival strategy. A tree which has suffered many injuries by storm, lightning, and man-made intervention has to restore its optimum design more frequently by adaptive growth than another tree comfortably protected by a dense stand. Just like the face of a restless adventurer tells more than that of an anxious stay-at-home, the tree which had to fight for its survival will always have the more inspiring appearance.

This book is meant to be a scientific but simple guide to interpreting the design of a tree in order to infer its individual mechanical load history. The reader is invited to contemplate and comprehend. He is taught how to read in the tree's biomechanical diary.

The knowledge presented was gained at the Karlsruhe Nuclear Research Centre during the last years through the use of modern computer methods such as the finite element method. A range of technical reports is available to readers wishing to acquire more insight into the compiled theoretical-mechanical results. The present book, however, addresses the forester and biologist, the tree expert and tree fan, the hiker and lover of nature who want to learn more about the trees encountered outdoors.

Since, in the final analysis, trees have taught us to effectively optimize engineering structures, this book is also intended to provide a biomechanical background for designers and engineers using the CAO method (Computer-Aided Optimization) to shape-optimize engineering components by letting them grow like trees. Details on the CAO method are given by Mattheck (1990 a) and Mattheck and Burkhardt (1990).

The contents of this book and related theoretical studies are the subject of a lecture on biomechanics, which has been held by the author for some years at the University of Karlsruhe.

The perhaps most fascinating fact about all the results presented is that, obviously, only two mechanical design rules govern the structural growth of the tree: *the hypothesis of constant stress on the surface of the tree and the law of external load minimization*. Both rules will be introduced to the reader and elucidated by various practical examples.

Throughout the book, the representation of trees by drawings has been preferred to photographs. Drawings allow one to omit the biomechanically irrelevant details, and make it easier for the reader to concentrate on the essential. All examples presented were drawn by the author according to actual examples taken from nature.

Karlsruhe, February 1991 — Claus Mattheck

CONTENTS

1. INTRODUCTION

There is nothing more uncompromising than competition in pure and unadulterated nature. Nowhere is the imperfect, non-optimized design hunted, attacked or threatened as pitilessly as in the difficult struggle for energy and living space to which animals and plants are exposed. No one could seriously doubt the hypothesis that competition leads to the selection of the most functional and adapted mechanical designs, i.e. the survival of shape-optimized biological load carriers which are as light as possible and as strong as necessary. Corroborating this hypothesis with regard to trees, it seems proper to start by giving a brief sketch of their biological features and functions.

Since this is a mechanical and not a biological study of trees, the main biological functions are only roughly and self-explanatorily outlined in Fig. 1. A similarly straightforward representation is given by Wilson (1984). All problems of tree structure are discussed from a more biological point of view by Zimmermann and Brown (1980), who also compile and examine a comprehensive bibliography.

Regarding aspects of mechanical relevance, the so-called secondary growth in the thickness direction of the tree parts is of extreme importance. It is the thickness of branches, stems, etc., which determines the mechanical stiffness, the supporting qualities of tree design. Secondary growth takes place just below the outer surface of the stem or branch in the cambium, where phloem, and finally bark, are produced outwards, and wood is grown inwards. The inwardly grown wood is an annual tree ring consisting mostly of more porous early wood, and denser and stronger late wood. Tree rings are arranged around branches, wounds, and inclusions following very sophisticated design rules. Being the inner architecture of the tree, they mainly determine its material properties and help the tree to survive.

As evident from Fig. 1, the design of the tree must necessarily be a compromise. Although it is a good photoreceptor, a large crown is heavy, and many supporting members (stem, branches) are required, which have to be nursed, maintained, and, if necessary, repaired. The best possible tree design is needed to minimize this material effort. A large crown loses much water by transpiration. The water evaporated must be replaced by sufficient quantities of fresh supplies

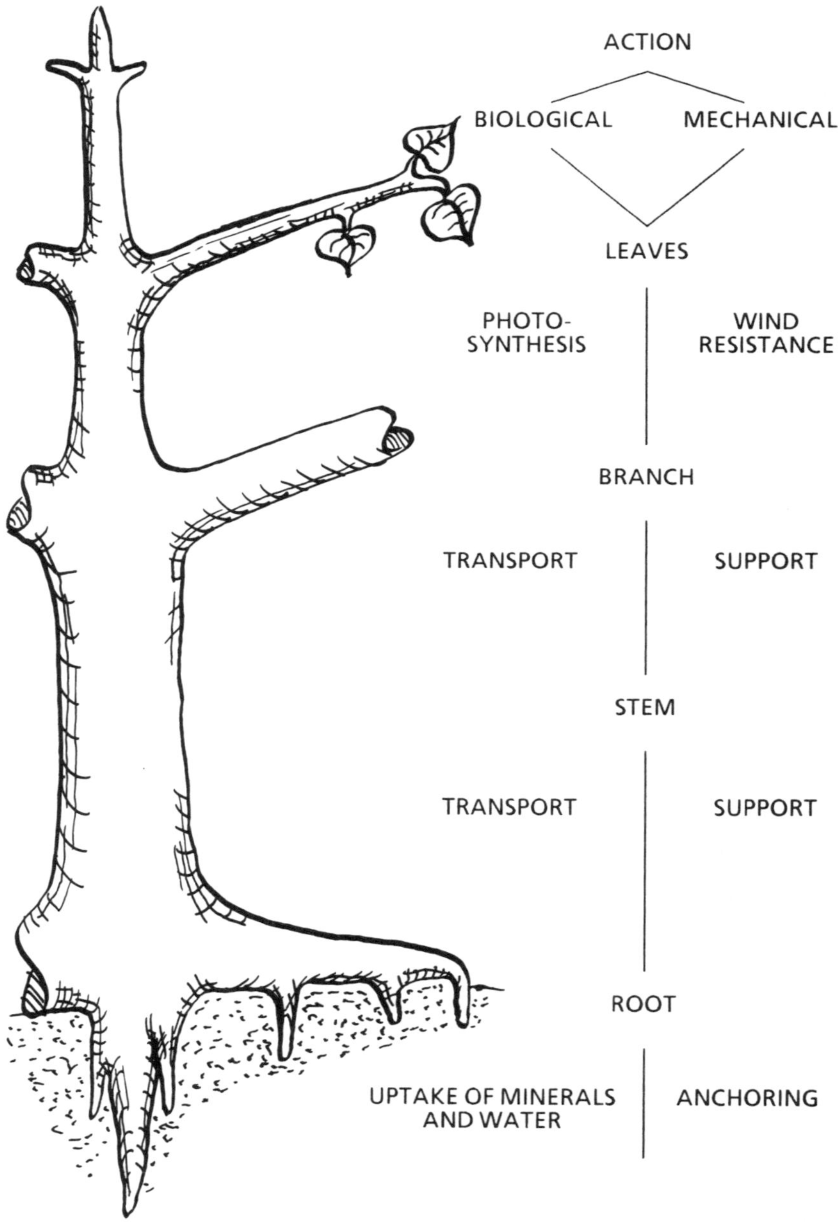

Figure 1: Major tree parts and their biological and mechanical functions

transported to the shoots by an equivalent root system. A large crown is more susceptible to wind attacks. The stem must be more resistant to bending, and the root system be anchored fast to resist wind loads, which means that still more material of sufficient strength is required, which again has to be nursed and maintained.

All these counteracting requirements result in the ideal "compromise tree" characterized by the largest possible crown but the smallest number of supporting members, all of which are of sufficient strength, but not so large as to add weight and increase energy consumption. The truly fine-tuned design of the "compromise tree" is based on a number of mechanical fundamentals with which anyone studying trees should be made familiar.

2. THE ABSOLUTE MINIMUM OF MECHANICAL KNOWLEDGE

2.1 EXTERNAL LOADING

Trees are exposed to many load cases. The wind, for example, varies in direction, and each direction is related to an individually adapted cross-section. Bending moments due to wind can be accompanied by other external loads such as torsion, or axial compression due to weight. The relevant load parameters are introduced in the following to the reader who is unfamiliar with them.

A force F acting on the end of a cantilever beam causes the beam to bend. An equal and opposite bending moment M_B is required at the clamped end of the beam to fix it against rotation (Fig. 2). Another force acting on a rectangular lever arm (Fig. 2) introduces a torsional moment into the clamping. While the longer part of the lever arm is rotated around its axis, its shorter part is loaded by bending alone (see example above). In addition to torsion, the longer part, too, is loaded by bending.

It would be beyond the scope of this book to go into details. The beam and lever examples shown in Fig. 2 illustrate the terms *bending moment* and *torsional moment*. Other forces acting on the clamping of the beams have been ignored,

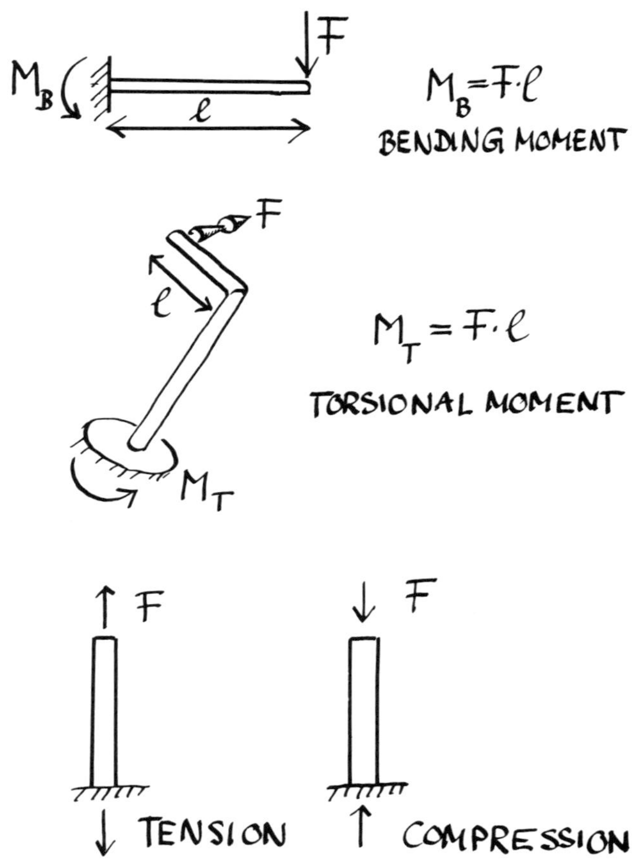

Figure 2: Classification of external loads into bending moments, torsional moments and axial forces

and plotted instead in the moment-free examples below, where the clamping has to counterbalance axial forces, i.e. tension or compression, respectively.

Obviously, the beams have the function of "messengers" transmitting external loads from one end into the clamping at the other end. To do this, they have to deform by bending, torsion, lengthening, or shortening. The process of deformation is completed as soon as the equilibrium of internal stresses and external loads is restored. The inner stress distribution near the restraint of the beam is its mechanical "message" transmitted into the clamping. There, the sum of all stresses is integrated and counterbalanced by equal and opposite reactions preventing the beam from canting, yielding, or subsiding. The stresses in the mechanical component have a well-defined inner distribution which depends on the nature of the external load applied, and on the type of clamping (Fig. 3).

2.2 INTERNAL STRESSES: THE RESPONSE OF THE MECHANICAL COMPONENT TO EXTERNAL LOADS

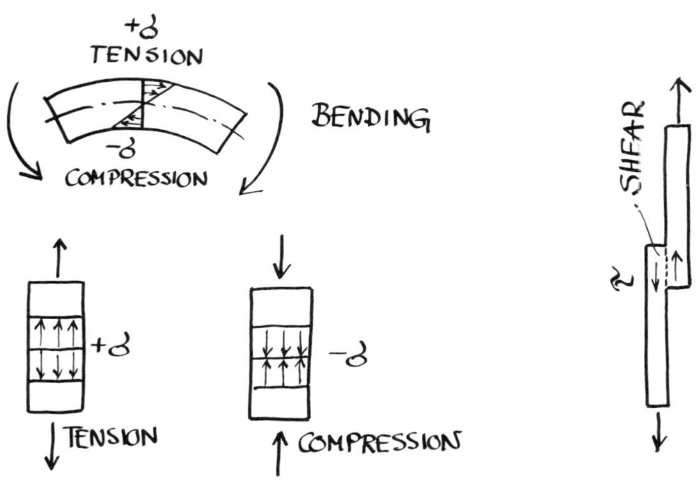

Figure 3: Simplified classification of externally induced internal stresses into tensile stresses, compressive stresses and shear stresses

In the case of bending, stresses change from tensile at the convex side to compressive at the concave side of a component. This type of loading is found, for example, in trees exposed to the wind. Axial tensile forces induce tensile stresses which may be distributed homogeneously over the cross-section of the load carrier. The same applies to compression forces. Different by sign only, they induce compressive stresses. At least in the stem bearing the weight of the crown, a tree is also loaded by axial compression, which, however, is not as significant as bending induced by wind loads. Internal, or growth stresses, produced by the living tree itself are also more important; they will be discussed in a later section.

While bending stresses are the component's internal resistance to flexure, tensile stresses resist lengthening. Compressive stresses limit shortening, and shear stresses prevent component parts from slipping on shear-loaded interfaces. All in all, stresses are the inner response of components to the loads applied externally. Mechanical components are destroyed if stresses reach critical values at any point of the structure. Such values are the material parameters which characterize their strength. They have to be determined experimentally. The stress and strength aspects referred to above lead up to a point of major relevance: The component will fail if a critical stress state is reached somewhere in its inner structure or on the surface. A good design is always characterized by homogeneous stresses, i.e. an even distribution of loads. Throughout such components, stresses are distributed equally, meaning that no point has to bear more than the other, and no point is predetermined to fail.

Just the opposite is true for poor designs. They are characterized by one or more typical points of failure causing components to break in service in the same way due to high localized stress peaks. This is especially true near notches, i.e. near any concave part leading to a diversion of the force flow. In most cases, technical notches will cause stress peaks which, for example, are at least three times as high as the stress applied in the case of a circular hole in a tensile plate (Fig. 4).

Figure 5 shows some selected notches. The dots indicate the expected locations of stress peaks, i.e. notch stresses. They are found in places where the density of the diverted force flow (arrows!) is very high, i.e. where the force flow must pass through a very small part of the cross-section.

Turning to biological structures, which are capable of adaptive growth, one cannot but be amazed at their achieving the unbelievable: *adaptively grown notches do not cause notch stresses*. Section 4 discusses the hypothesis of constant stress on the surface of the tree, and gives a more detailed account of this phenomenon.

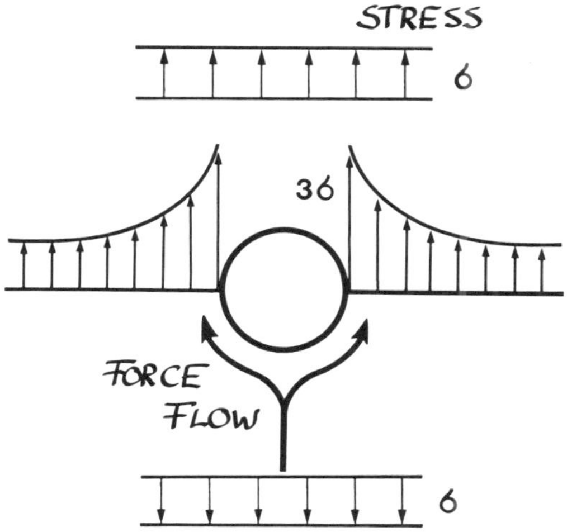

Figure 4: Externally applied tensile loads are at least tripled in the presence of a circular notch (hole) in a plate because the force flow has to be diverted around the hole.

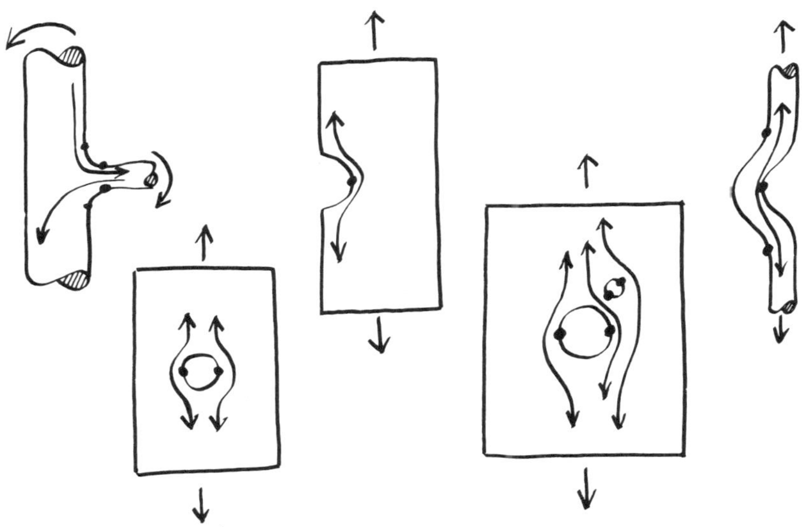

Figure 5: Any force flow diversion (arrows) can lead to a high local increase in stresses (stress peaks). The dots indicate the expected locations of stress peaks.

The above simplified technomechanical considerations help to classify the types of external loading to which trees are exposed, namely: axial forces (tension, compression), bending moments, and torsional moments. The tree reacts to external loading by internal stresses subdivided roughly into tension, compression, and shear stresses. Bending stresses are a combination of tensile and compressive stresses acting in a well-defined distribution over the cross-section. Since tree failure starts as internal stresses take on critical values as a result of external loads, an advantageous shape, which minimizes all avoidable external loading, is of vital importance to the survival of any tree.

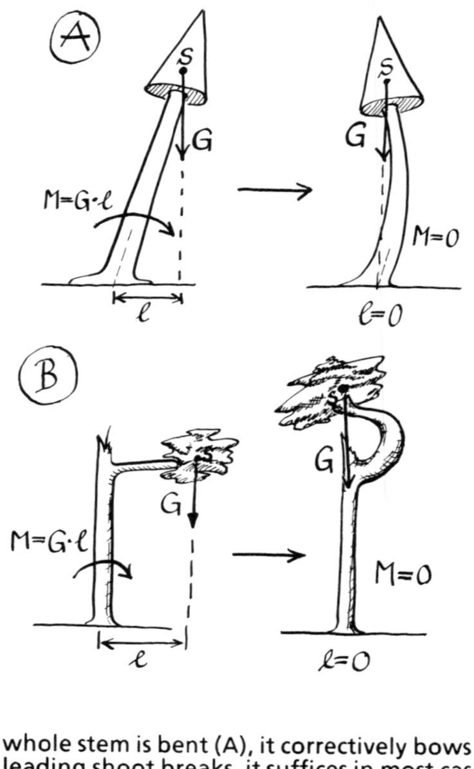

Figure 6: If the whole stem is bent (A), it correctively bows along its length. If the leading shoot breaks, it suffices in most cases to put the replacing side branch upright (B) and rearrange the centre of gravity of the crown above the root stock.

3. HOW THE TREE MINIMIZES ITS EXTERNAL LOADING

The two sketches on the left in Fig. 6 show trees whose butts are unnecessarily loaded by bending, because the centres of gravity of the crowns are not over the root stocks.

However, since trees have receptors, which respond with a corrective mechanism as soon as they perceive wrong tendencies or bends, they adapt and develop into the improved designs shown in the figure. In Fig. 6, bending moments are avoided at the bases, and are definitely reduced in the upper sections of the stems. "Negative geotropism" or "negative gravitropism" are the terms describing this self-correcting, countergravity growth phenomenon.

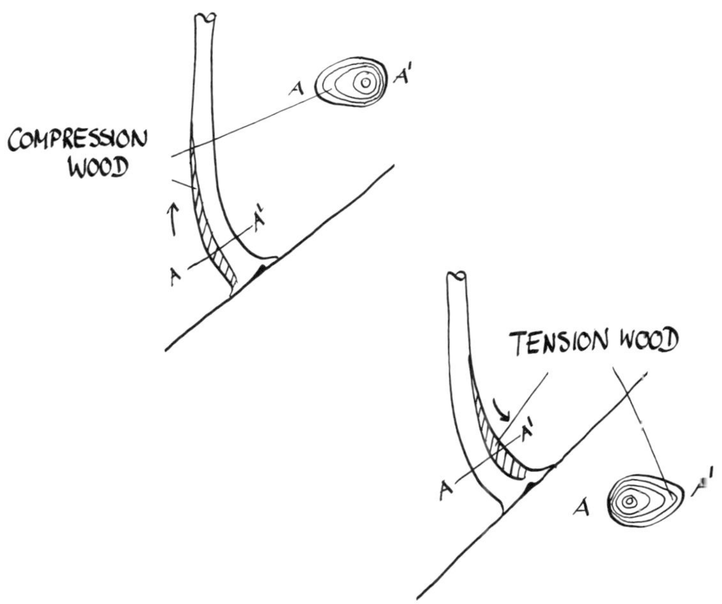

Figure 7: Gymnosperms are pushed upward into the vertical position by formation of compression wood at their lower sides facing the soil. Angiosperms are pulled into position by formation of tension wood at their upper sides facing away from the soil.

Considering the forces behind such powerful, upward self-deformations and dynamic internal stresses, which correct the orientation of stem or branch, one comes up with the principle outlined in Fig. 7. Two examples were chosen to explain it on the basis of the fundamentally different behaviours of deciduous (angiosperms) and coniferous trees (gymnosperms).

The gymnosperm (Fig. 7, above) forms *compression wood* at its lower side facing the soil. The tree is *pushed* upward into the right position by axial compression wood expansion. This process can be compared with heat expansion, i.e. upward bending as the reaction to heating at the lower side of a material. The same corrective forces are illustrated for a slanted angiosperm (Fig. 7, below), which forms *tension wood* at its upper side facing away from the soil. In this case, the tree is *pulled* into position like some material cooled down at its upper side. Of course, there are no thermal stresses inside the real tree. This comparison is simply used in order to illustrate growth stresses.

The push-and-pull responses described above are referred to as *reaction wood* (tension and compression wood), which, so to speak, is the executive part of the negative-geotropism growth regulator. Figure 8 shows all the major regulators controlling the growth of trees.

Phototropism directs the tree towards the light. It is a light-intensity indicator controlling the growth of trees according to the light conditions.

Apical dominance, in the sense of growth regulation, is the manifestation of the dominance of the leading shoot. It is a distance-demanding order intended to keep the side shoots away from the leading shoot. The executive mechanism, i.e. the increased growth of upper branch sides, is called epinasty.

It is evident from Fig. 8 that growth regulators can be conflicting. Both geotropism and phototropism cause the branches to grow upwards. However, all non-leading, subordinate branches are suppressed epinastically by the leader's dominance signal to take a horizontal or even downward position.

The optimum compromise of the branch angle achieved in this way may change with increasing distance of branches from both the stem and the soil.

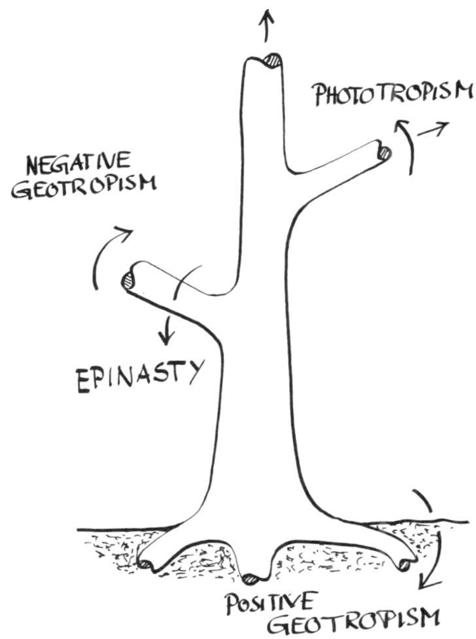

Figure 8: Major regulators controlling the growth of trees in a both conflicting and concurrent way.

Evidently, trees are able to avoid or to minimize unnecessary external loads. However, loads such as wind, their own weight, and internal growth stresses still have to be carried, and significant internal bending stresses caused by the very formation of corrective reaction wood must be endured in order to survive.

While Section 2.2 pointed out that notch stresses can be avoided, and that good design is characterized by an even, i.e. a constant, stress distribution on the tree surface, the following explains how this can be done and maintained in practice.

4. EVEN LOAD DISTRIBUTION: THE HYPOTHESIS OF CONSTANT STRESS ON THE SURFACE

Figure 9 presents the main sources and effects of mechanical stresses acting in trees. Emphasis will be on axial stresses, since lateral stresses account for only about one-tenth in comparison.

Stresses are highest where bending (wind loads) and growth stresses (Kübler 1959) are involved. The latter will be discussed in detail together with the end-splitting phenomenon.

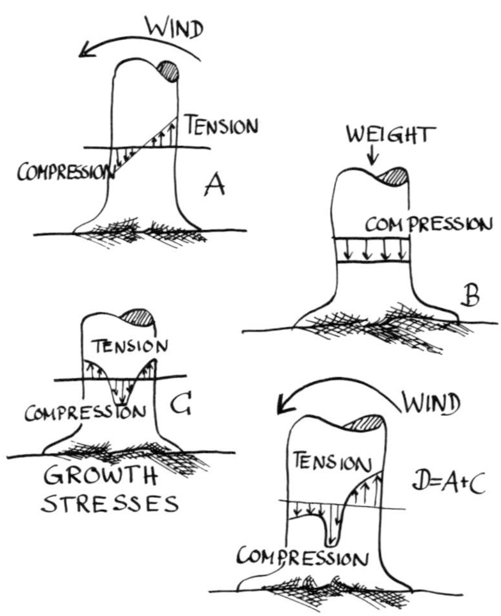

Figure 9: Main sources and effects of mechanical stresses acting in trees

According to Kübler (1959 a, b), superimposed growth and wind bending stresses slightly reduce the critical compressive stresses, which in stormy weather may lead to fibre buckling at the compression side of bending. Thus, the superposition of stress types A and C (Fig. 9) shows stress states of type D, which are characterized by slightly increased tensile stresses. Being the first to derive an analytical description of growth stresses, Kübler (1987) also wrote an excellent, detailed overview, which outlines the state of growth stress research up to 1987.

Stresses caused by the tree's own weight exist, but are not as significant as others. This can be concluded from the fact that, while trees often break under wind loads, and may end-split after felling due to growth stresses, no healthy tree was ever seen to buckle under its own weight. Much to the surprise of the author, failure by buckling has nevertheless been discussed by McMahon (1973), and comparisons have been made between measured height-diameter relations and relations calculated from Euler's buckling theory.

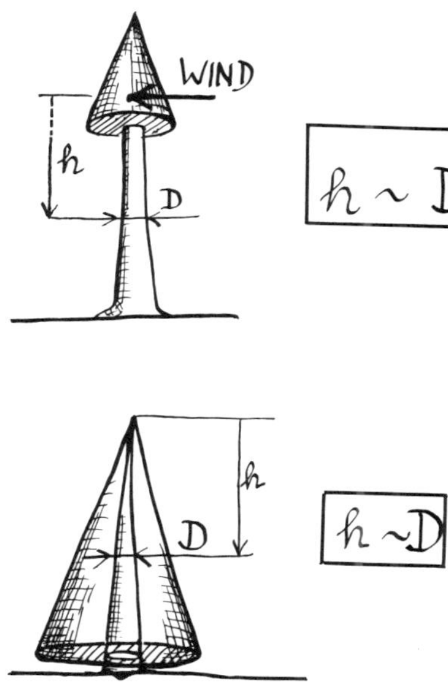

Figure 10: The relation between local stem diameter D and the respective height h is essentially determined by the wind load. This accounts for the importance of crown forms and sizes.

Metzger (1893) has shown that the height-diameter relation of spruce trees obeys the law $h \sim D^3$ with excellent accuracy. It is this very relation which, assuming a constant axial surface bending stress along the length of the beam, can be calculated theoretically for a constrained bending bar (cantilever beam) loaded by a lateral force (wind load) at its free end. In accordance with this hypothesis of *constant mechanical stress* along the length of the stem, spruce stems are bending bars with axially constant stresses on their stem surfaces. They taper off towards the top to adapt themselves to the decreased wind loads higher up (Fig. 10).

Of course, there is a difference between small crowns localizing wind loads in the very top (Fig. 10, above), and larger, overall crowns (Fig. 10, below).

Theoretical proof can be given of the fact that the latter case of a linearly downward increasing crown width leads to a linear height-diameter relation $h \sim D$ if constant stress along the stem is postulated. This has been verified by the author on a small scale for some selected Leyland cypresses. Observations were found to be in excellent agreement with the calculated results.

In accordance with the above, the relation between stem diameter and height is defined by the postulate of constant axial stress on the stem surface of a tree.

The compiled case studies will illustrate that branch joints, root joints, trunk weldings, etc. are shape-optimized, too, to completely avoid notch stresses, and to establish a constant stress state on the surface of the "mechanical component tree". This has been confirmed at the Karlsruhe Nuclear Research Centre by numerous finite element calculations by Mattheck (1988, 1990 a, b).

Trees dedicate their lives, so to speak, to the preservation of their most optimum design. They maintain and, if necessary, restore the *state of constant stress on their surface* by permanent adaptation to the everchanging external loads to which they are exposed.

Among the typical examples of stress-controlled healing and self-repair are wounds caused by mechanical impacts.

From the mechanical point of view, such wounds are notches leading to a diversion of force flows, and to local stress peaks (notch stresses). Theoretical investigations by Mattheck and Korseska (1989) have shown that such wounds (notches) heal fastest where stress peaks are highest due to force flow diversion, i.e. at the contour points which deviate most from the ideal constant stress state.

Failure is most likely to occur where notch stresses are highest. Any such weak spot in the tree design is given preferential care, and is repaired in the quickest possible way. Figure 11 shows a branch hole with very pronounced, local wound healing characteristics (callus formation) at side points where notch stresses are highest.

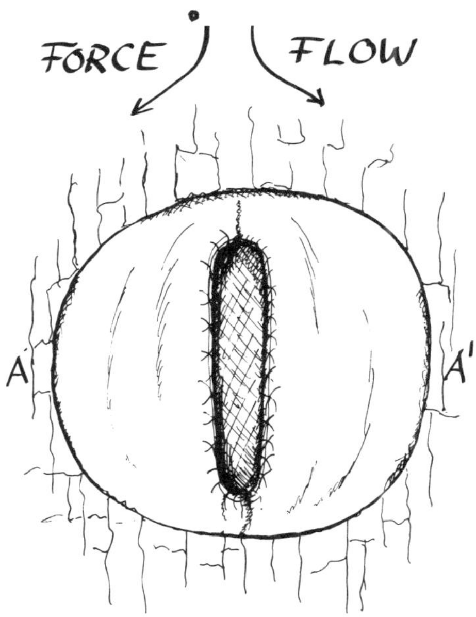

Figure 11: Branch holes heal into vertical spindles because wound healing is stimulated mechanically by the very high notch stresses at side points A, A'.

Restorative action is also taken if a foreign body, e.g. a stone, a dead branch from the same tree, or a tree alien to the species, keeps pressing against the surface of the stem (see case studies). The high contact stresses (Hertzian stresses) induced are another type of disturbance of the tree's ideal constant stress state.

Figure 12 shows a stone which rolled downhill, was stopped by a tree, and was pressed against it by downslope forces. In this case, the tree reduces the resultant contact stresses by enlarging the contact surface. Growing around it, the tree envelops the stone until, at least in some cases, it is completely integrated.

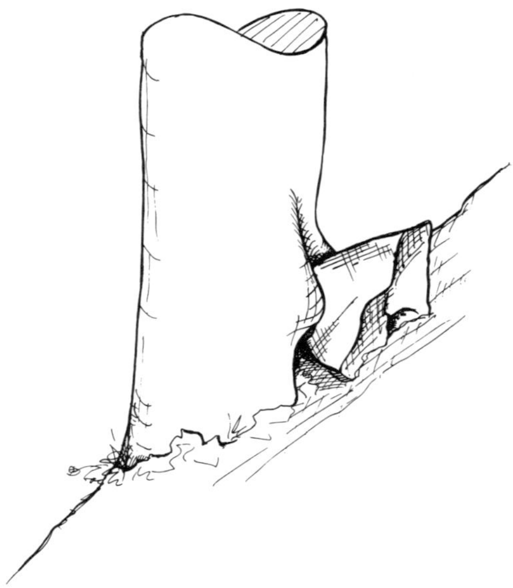

Figure 12: A stone contacts a tree. The tree reduces the contact stresses by growing around the stone and enlarging the contact surface.

The final purpose of these adaptive growth activities is the best possible restoration of the constant stress state, i.e. an even distribution of mechanical stresses.

The different mechanical optimization processes, the restorative mechanisms and repair patterns described above will be outlined again and defined in an axiomatic way in the following.

5. MECHANICAL CLASSIFICATION OF ADAPTIVE GROWTH

Adaptive growth can be defined as the elaborate shaping of individual tree designs well-adapted to external load conditions. No corrections or modifications are made unless external circumstances demand that they be made. Trees remain intact with no need for restoration as long as they are left uninjured and undisturbed by impacts, notches, or contacting bodies. Only if the optimum state of constant stress is in some way disturbed, will adaptive growth restore the ideal tree design by adapting to the new situation. No difference is made here between constructive and restorative growth, because, being equally important, similar types of adaptive growth, both serve to develop the best possible tree designs in accordance with the respective prevailing conditions. The life of a tree can be characterized as a constant "body-care" effort, which for a long time can even compensate inner decay by the external addition of material. The tree dies once its adaptive growth abilities begin to fail and reactions become too slow to compensate the rapidly changing external conditions. The following diagram elucidates the broad spectrum of adaptive growth activities and mechanisms:

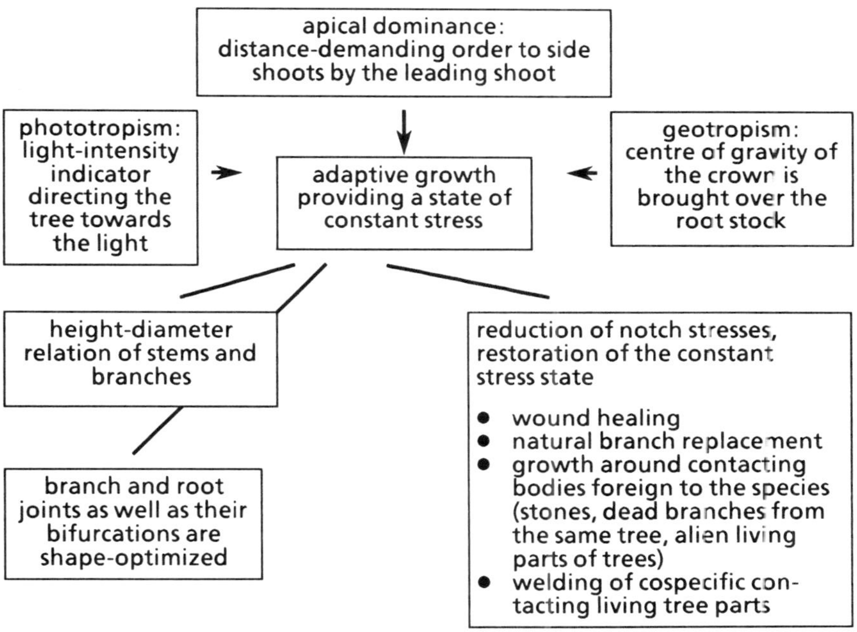

The arrows coming from apical dominance, phototropism and geotropism indicate that further requirements stipulated by these regulators must be met by the tree in addition to the constant stress requirement.

If, for example, negative geotropism demands that the orientation of a branch or a stem be changed, reaction wood is formed. Additional growth stresses are the natural consequence, but moments of external bending and torsion are minimized in this way, and finally relieve the tree in its adaptive efforts.

Phototropism, for its part, is quite often found to be balanced by geotropism. Any stem or branch stimulated to prefer a particular growth direction due to locally incident beams of light is controlled in bending by the geotropic function which counteracts slanting and bending due to its own weight. The newly adopted compromise position of equilibrium induces higher bending loads, and adaptively increases the girth of the tree until it has regained its optimum shape which enables it to cope with the new situation.

The beholder will find it feasible to trace the history of a tree, because, unlike bones, wood does not shrink under reduced loads. Wood is attached at overloaded places instead, but is nowhere removed or shed unless through decay or mechanical destruction. The coexistence of the stump of a dead leading shoot with a successor above it is due to this preserving property.

Preserved material and parts as well as the unconditional, continued existence of older tree rings facilitate grasping the fates of most trees. The biomechanical laws set out below can be a helpful tool in this endeavour.

1. Geotropism

In the case of evenly distributed light, trees minimize their bending due to gravity and torsional loads (wind loads) by placing the centre of gravity directly over the centre of the root stock.

2. Phototropism

If the incidence of light is only from one direction, ideal geotropic growth is hampered. A compromise design providing sufficient light reception but avoiding excessive external loading (for example, through a slanted tree's own weight) is then adopted. Higher bending stresses incurred as a result are compensated by a suitable adaptive increase in girth.

3. Apical dominance

The leading shoot asserts a claim to apical dominance. Its distance-demanding signal suppresses the side branches by epinasty, i.e. more intense growth of the upper branch sides, and counterbalances the upright message of negative geotropism.

4. Adaptive growth provides a constant stress state on the surface of the tree

- by adjustment of an optimum height-diameter relation of stem and branches,

- by shape optimization of branch joints, root joints, and bifurcations to avoid notch stresses,

- by stress-controlled healing of wounds (e.g. notches accompanied by notch stresses!) caused by mechanical impacts, i.e. maximum wound healing in places of highest stress,

- by reduction of mechanical contact stresses caused by local body contacts. The contacting surface is enlarged as the tree grows around the body to envelop and finally integrate it. Bodies foreign to the tree are, e.g. dead objects such as stones or scraps of iron, their own dead branches, and dead and living branches alien to the species.

- by welding the contacting parts of trees of the same species, i.e. formation of common cambium and tree rings after unification of the contacting parts. In most cases, this leads to reduced and more advantageous load distributions, modified height-diameter relations, and less consumption of material (framework principle),

- by arrangement of tree rings along the path of the force flow, i.e. along the principal stress trajectories, and thus along the shear-free inner parts of the component. Dangerous shear stresses between the tree rings as well as bending of the individual tree ring are minimized in this way.

Geotropism, phototropism, apical dominance, and adaptive growth are the major growth-regulating mechanisms to be taken into account when studying trees. Instead of discussing the whole range of other -nasty- or -tropism-suffixed regulators contributing their share, we prefer a substantiating, commentated presentation of some selected case studies.

The reader should note that the constant stress principle only applies to the actually load-bearing, structural members of a tree. For example, an old stump of a nearly decayed branch is not particularly loaded, and, therefore, cannot be expected to have the same stress level as an intact, loaded branch. Unlike bones, trees do not shrink. Unless they break off or are removed as a result of mechanical impacts, non-loaded (i.e. unneeded !) parts remain members which can only be inactivated by interruption of nutrient transport, death, and decay. This mechanism can be understood as a type of rudimentary and very simple passive "shrinkage". Since trees, unlike animals, do not move about carrying their useless burdens, this primitive kind of shrinkage suffices to eliminate even the heaviest non-loaded structural members.

6. CASE STUDIES

The design of a tree is a living record, like the human face, showing the signs of life. Since it is difficult to find two exactly identical fates, one can hardly expect to find two completely identical tree designs. It is unreasonable, therefore, to judge and evaluate trees by stereotype criteria. Judging trees is a difficult task which certainly needs some practice. It is not easy, for example, to determine the actually load-carrying parts, i.e. the recently added parts of a tree. Non-carrying parts may still exist because, as mentioned above, trees do not shrink like bones. However, they do reveal earlier load situations, i.e. the mechanical history of the tree. The following examples will enable the reader to increase his awareness of individual details with a good eye for the essentials. All examples, taken from nature, are self-explanatory with detailed descriptions added to each.

TROUBLE-FREE CHANGE IN LEADERSHIP

The leading shoot of a conifer becomes diseased and dies. There is only one possible successor which once may have been as small in diameter as the tiny dead side branch above it. Conifers take an upright position by forming compression wood (geotropism) on their lower sides. This diversion mechanism is stopped first at the very top of the tree, where bending in the opposite direction starts as soon as the lower side changes from the right to the left side of the branch. As soon as the centre of gravity of the crown is situated once again over the butt of the stem, the mechanism is inactivated, and the new leading shoot takes its place. Result: Minimized external bending moment due to crown weight especially at the butt of the tree, as well as reduced torsional moment in the case of wind blowing perpendicularly to the plane of the page. There is virtually no torsional moment acting if the wind blows in the plane of the page. An obviously unavoidable moment of bending due to crown weight is still acting in the knee formed at the base of the diverted branch.

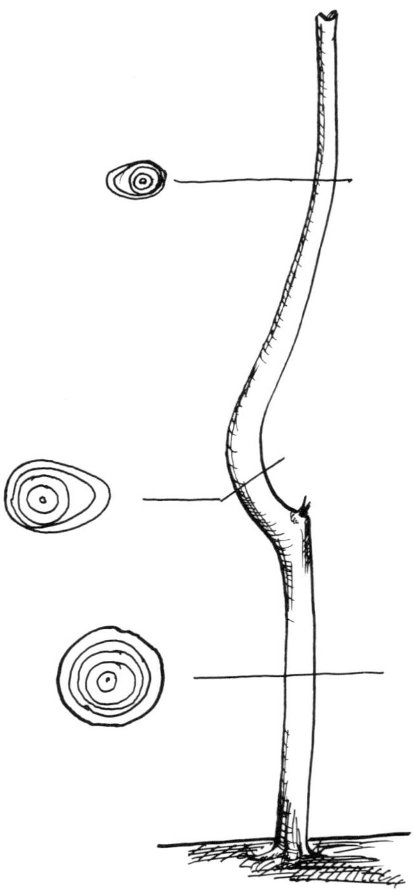

TROUBLE-FREE CHANGE IN LEADERSHIP IN A BEECH TREE

This beech is another example of a trouble-free change in leadership. Enabling the tree to successfully compete with the neighbouring trees, the replacement of the leading shoot was to its own benefit. The beech as an angiosperm (deciduous tree) takes an upright position by forming tension wood at the upper side of the slanted branch, i.e. the right-hand side of the stump of the old leader. In the tree top, tension wood is formed first on the right- and then on the left-hand side of the stem to restore the vertical position. By sawing through the stem of a tree, one can determine the year of death of its leading shoot by the last of the tree rings of nearly circular shape. Once the shape of the tree is corrected, the widths of all subsequent tree rings are assumed to be equal again. Note that this does not necessarily imply roundness, but only equal thickness of the outermost tree ring. The dry, small, remaining stump of the former leader will soon completely decay and disappear. After closure of the collar-surrounded wound only a small hump will give evidence of the former leader.

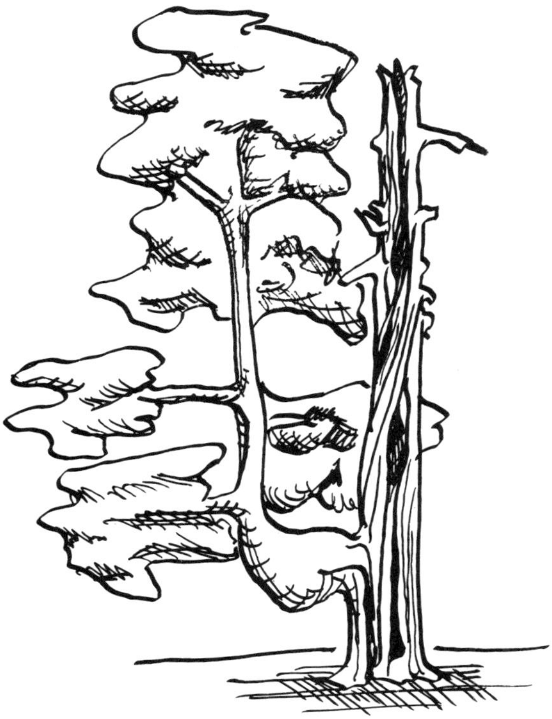

A LINGERING LEADING SHOOT
(after a photograph by Dr. F.G. Buchholz)

Even old conifers are capable of replacing dead leading shoots. However, in the present case, the ascending side branch is impeded by the presence of the old stem which, although dead, is still quite stable. The centre of gravity cannot be simply moved over the centre of the root stock. Knowing that there will be no epinastic distance-demanding signals from the old, dead stem, and presupposing that the living branch will sooner or later be pushed upright by its own geotropism, one may wonder why the branch does not approach the old, dead stem any further. There are two possible reasons: First, the branch is at the southern side, and approaching the stem would mean growing into the shade. Second, the base of the branch has grown so thick that it is too stiff for further upright bending, and could not even bend if more reaction wood was formed there. This is perhaps the most plausible explanation, because the more flexible parts of the branch are already in a vertical position, and, therefore, do not form further reaction wood.

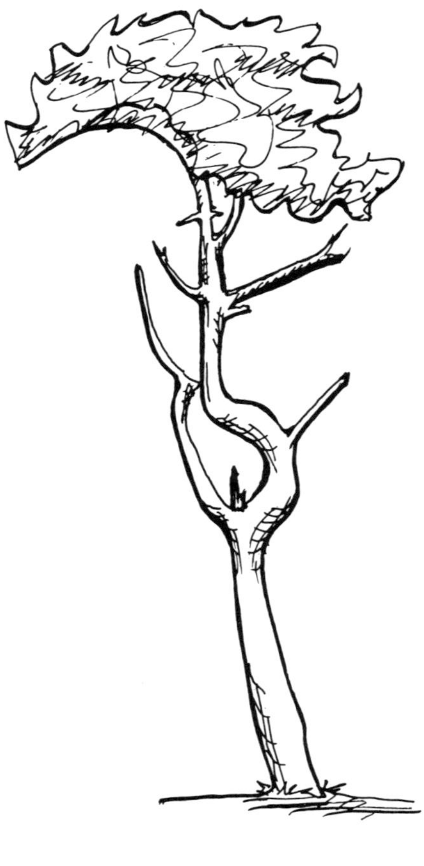

COMPETITION FOR LEADERSHIP

The leading shoot of a pine is damaged, weakened, and dies. It stops issuing its epinastic signal for there is no further need for the side branches to keep at a distance. Once epinasty fails to control the branches, competition for leadership starts, and each of the side branches produces reactive compression wood in order to take an upright position. Being on the sunny side, the right branch finally succeeds in being the first to place itself over the root centre, and thus restore the minimized bending of the stem. The new leading shoot starts sending dominance signals to the loser, which turns to the left and keeps its distance. The considerable increase in girth at the base of the new leader is due to formation of reaction wood, which in the case of this coniferous tree is compression wood at the lower side of the branch joint.

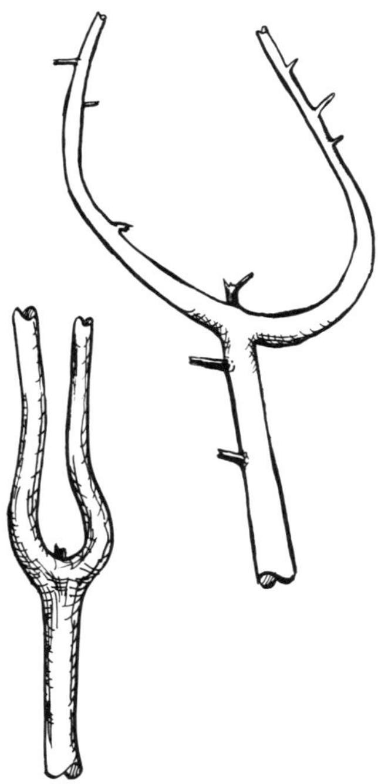

NONE OF THE SIDE BRANCHES LEAD

In the case of the pine tree shown on the right, the leading shoot also died. In the competition for leadership of two equal side branches growing under identical conditions, neither won or lost. They impede one another by sending dominance signals. The more they compete, the weaker they grow, and much material and energy are wasted before the tree nears the height of its neighbours. Facing windward, trees of that kind have a larger area exposed to attack, which leads to additional bending and torsional loads in the fork of the branches. It follows that two leading shoots are nearly as harmful as none. While the wide open fork of the pine is indicative of a position in the top of a tall tree, the narrow beech fork shown on the left reveals that under similar competitive conditions, the beech was situated very low in the stand among other trees, and was forced to narrow its fork in order to reach the light. Since two upward growing branches consume more material and energy than one, the narrow fork, too, is handicapped nevertheless by its two equal side branches.

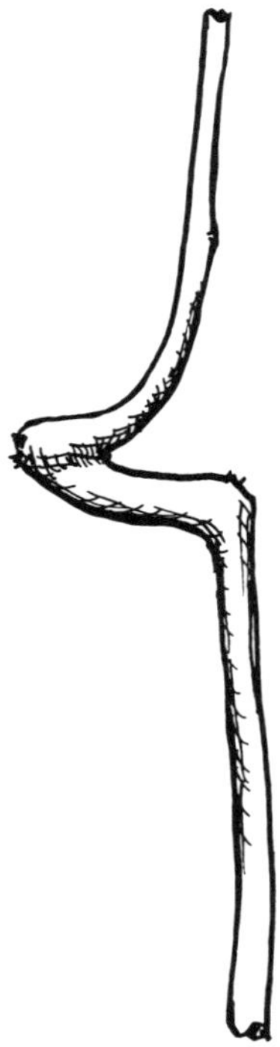

DOUBLE FAILURE AND CHANGE IN LEADERSHIP IN A BEECH TREE

The top of a thin beech stem breaks off. At the same time or shortly after, a side shoot breaks off, and loses its opportunity to rise. Originating from this side branch, a small shoot pointing upwards is induced by geotropism to grow and replace the crown's centre of gravity over the root centre. The tiny tree is just about 4 cm thick. It was obviously kept in a dense stand, which is why phototropism also played an important role in shaping its final design. Due to the presence of reaction wood and higher bending stresses, the horizontal part of the stem is thicker than the lower part. It is evident from this example that trees do not necessarily have to be age-old and timeworn to have an interesting shape.

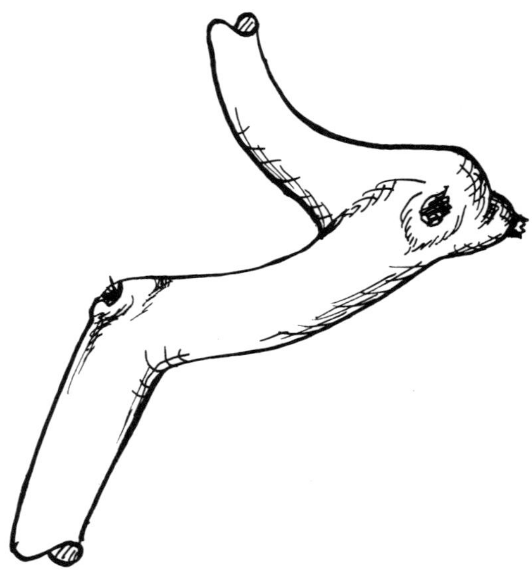

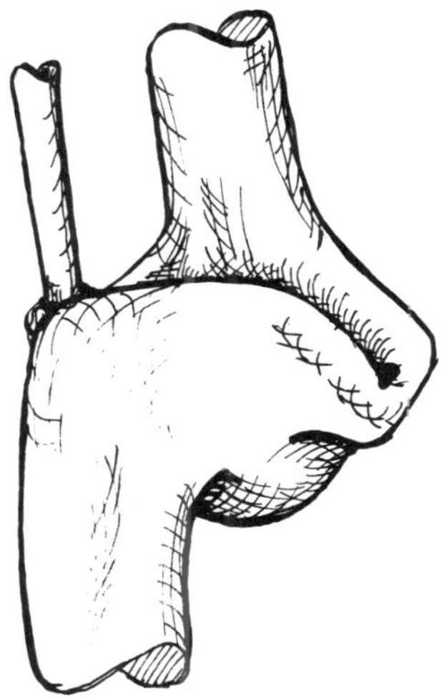

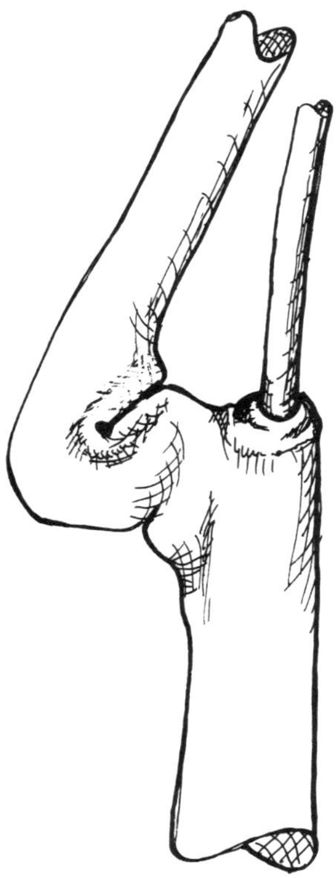

CHANGE IN LEADERSHIP ON THE VERGE OF SELF-DESTRUCTION

This angiosperm lost the leading shoot, which appears somewhat eccentric in the drawing. Induced by geotropism, the adjacent side shoot makes a sharp turn, which, as is typical of angiosperms, is achieved by formation of tension wood at the upper side of the branch. It must be the combined effect of reorientation and ceaseless, forced production of tension wood into the concave knee, which rapidly reduces the bending radius, and finally escalates in a sharp kink. Although tension wood is meant to place the side branch upright, it has an opposing action in this case because it is in the wrong place. The faces of the kink are squashed. Subsequent bulging of tension wood is due to contact stress reduction efforts, i.e. enlargement of the contact surface. The upper contacting edges will probably weld and defuse the situation. Such sharp kinks can occur if the branch to be diverted is still thin, and if, for example in dense stands, geotropism and phototropism are acting in the same direction.

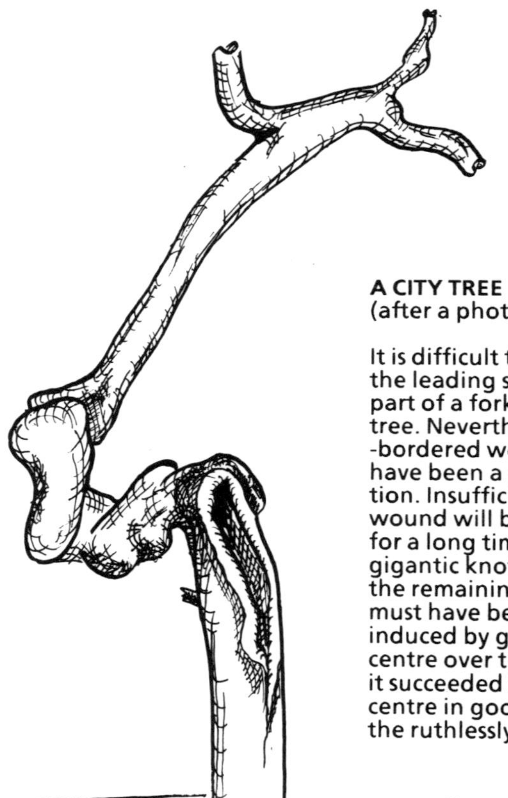

A CITY TREE
(after a photograph by J. Moeckel)

It is difficult to say whether it was the leading shoot or rather the right part of a fork which broke off this tree. Nevertheless, the large callus--bordered wound reveals that it must have been a rude and careless deprivation. Insufficiently repaired as it is, the wound will be inviting rot and insects for a long time. As is evident from its gigantic knots and awkward twists, the remaining left branch already must have been quite mature when induced by geotropism to place its centre over the butt. Nevertheless, it succeeded in situating the crown centre in good approximation over the ruthlessly incarcerated root centre.

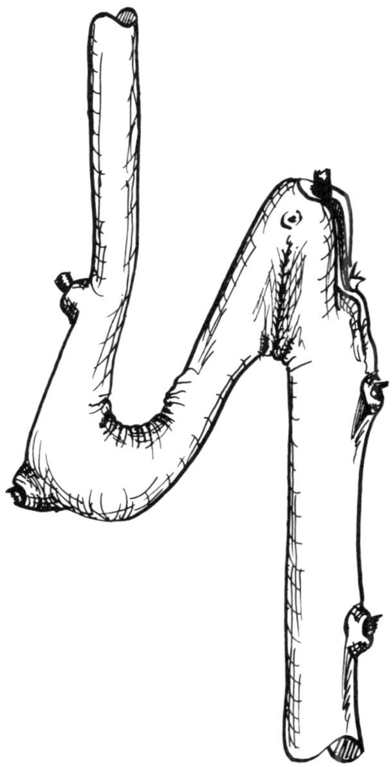

A YOUNG BEECH BRAVING ADVERSITY

This young, slender beech has grown toward the light from the very bottom of an extremely dark and dense stand. Its leading shoot was partially broken by snow, wind, or man-made impacts. One can tell by the annual rings of this tree that it was bent when only 2 cm thick. However, since at the concave (lower) side of bending some continuous wood fibres and the cambium remained intact, it was not broken completely. In spite of adverse growing conditions, the tree managed to heal the wound, and put the hanging, injured shoot or one of its side branches upright. Again, due to the combined geophototropic stimuli, the branch rises toward sun light. The heavily increased girth of the bending knee indicates that ample tension wood was formed to pull the branch upward. However, corrections of such acuteness are rare, and, if ever, are made by thin stems only.

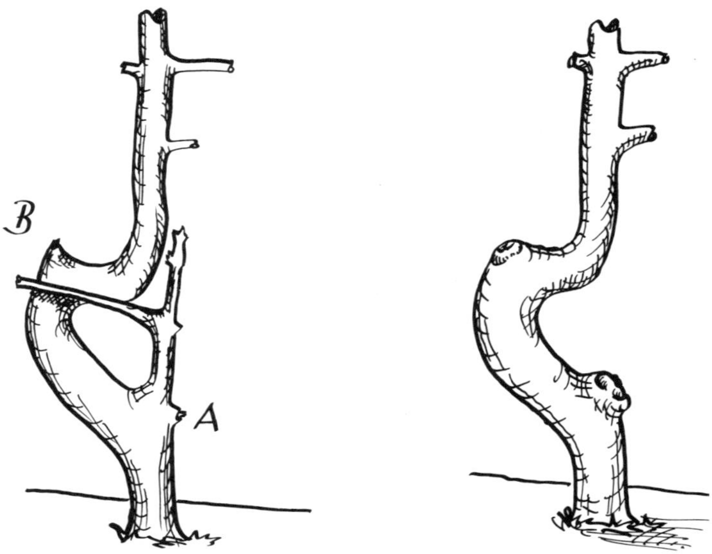

LEADERSHIP LOST TWICE

This pine tree standing on a beach was the first to make the author contemplate and study the physiognomics of forest trees. Already in its early stages of development, a pushing side branch tried to take the place of the stunting leading shoot (A). Through intense formation of compression wood at its lower side, the branch finally succeeded in putting itself more and more upright. When broken again in the midst of rising (B) by the heavy storms of the Atlantic, another tiny side shoot close to it made a sharp corrective turn, and placed itself perfectly above the root centre. High bending and torsional stresses due to wind loads contributed to fast healing at (B). The small stump which is still visible will soon be covered in the progress of further wound healing. Sooner or later the old vertical stump of the first leader (A) will probably also decay and break off. After that wound has healed, the tree will resemble the figure on the right. This indicates that it becomes increasingly difficult to trace the fate of a tree the older it is. This pine is a highly interesting tree which successfully withstood all attacks upon its life. Its design will continue to stop passersby, who will puzzle over it as long as it stays rooted in the beach.

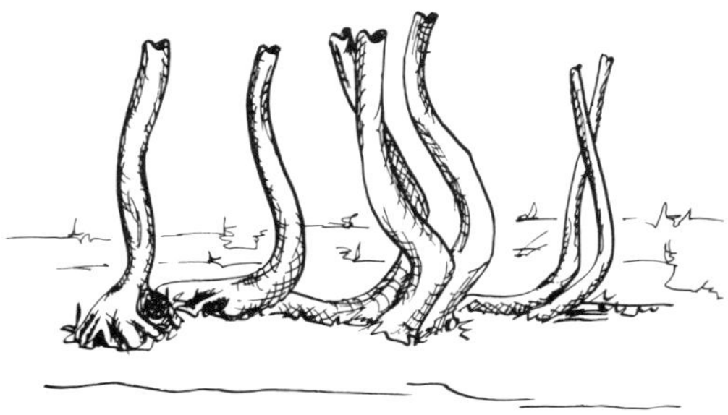

THE DANCE OF THE WAR BEECHES
(Professor J. Currey kindly took me to see these trees near Warwick, UK)

These groups of swaying beeches remind the observer of fairies dancing in a clearing. Their twisted shape was probably achieved by intervention of a gardener who notched their bases and kept them down for some time. They started to form tension wood at their upper sides so that they rose when released, and placed their centres of gravity over the centre of the root stock again (Currey pers. comm. 1990). In former times, Flemish tribes used such trees as a biological barricade against the troops of Julius Caesar, and indeed, such a wavy tree will block a wider area against an enemy cavalry than a straight tree in an equally dense stand of trees. A troop of experienced archers placed in addition behind such a barrier would repell even the best prepared attack on horseback. Of course, the beeches here are too young to really have obstructed a cavalry. Instead, it is very probable that a well-read gardener imitated this old trick to grow an interesting 100-m-long group of trees along the road. This impressive ensemble combines the silent majesty of trees with the swaying movement of S-curved lines in an ingenious way. It is a gift to the community of tree fans which has probably survived its creator.

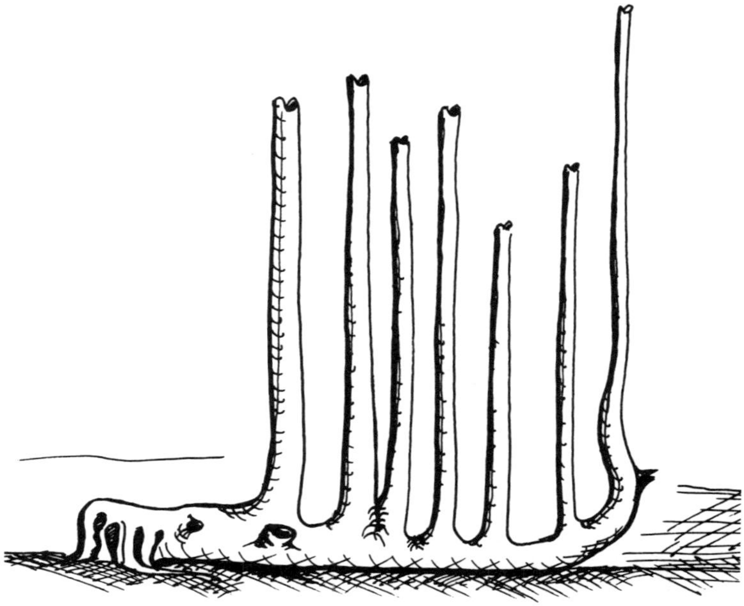

A PERSEVERING SPRUCE

The picture shows a spruce with half its root stock pulled out of the earth. The spruce must have had a certain age already when snow or wind succeeded in forcing it to the ground. Raising the whole stem could not have been done without a major mechanical effort (excessive formation of reaction wood), and the remaining anchoring root stock probably would not have been strong enough for such a long lever arm. Thus, the stem remained in a lying position, and only the flexible tip on the right was put upright through negative geotropic growth. Once the tree had assumed this position, all side branches grew into stems. An examination of new roots possibly grown along the bottom side of the stem or parts of it would answer the question whether the newly produced fibres are vertical or horizontal, i.e. whether the force flow enters the individual stems vertically, which is very probable, or whether high torsional loads are transmitted axially along the lying stem into the old root stock on the left, which is extremely improbable.

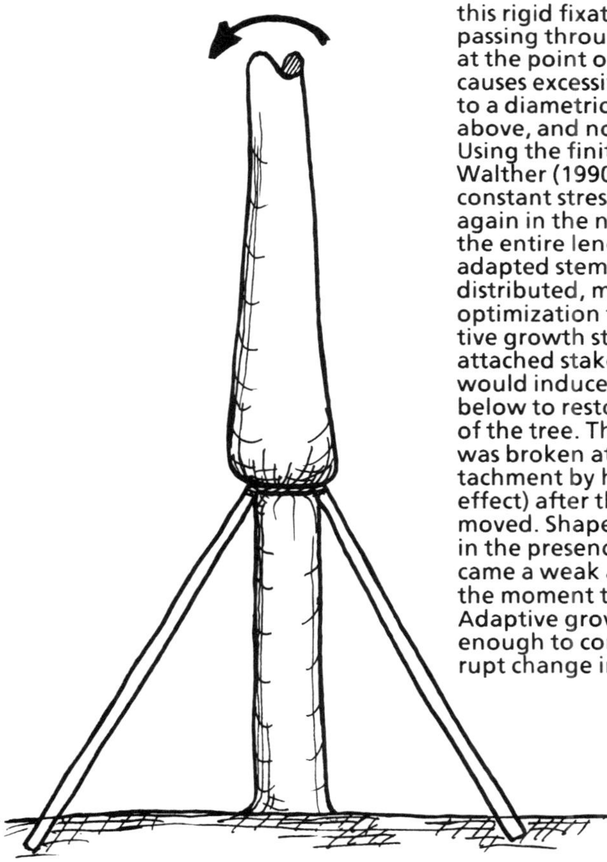

DIAMETRICAL ADAPTATIONS

To reduce wind loads, this conifer (*Leyland cypress*) was attached to supporting stakes after planting. While its lower part is relieved by this rigid fixation, the force flow passing through enters the stem at the point of attachment, and causes excessive stresses leading to a diametrical increase just above, and no increase below it. Using the finite element method, Walther (1990) has shown that the constant stress state is restored again in the new design. Along the entire length of the newly adapted stem, stresses are equally distributed, meaning that shape optimization took place by adaptive growth stimulated by the attached stakes. Stake removal would induce diametrical growth below to restore the normal shape of the tree. This conifer, however, was broken at the point of attachment by heavy storms (notch effect) after the stakes were removed. Shape-optimized as it was in the presence of stakes, it became a weak and faulty design the moment they were removed. Adaptive growth was not fast enough to compensate the abrupt change in loading conditions.

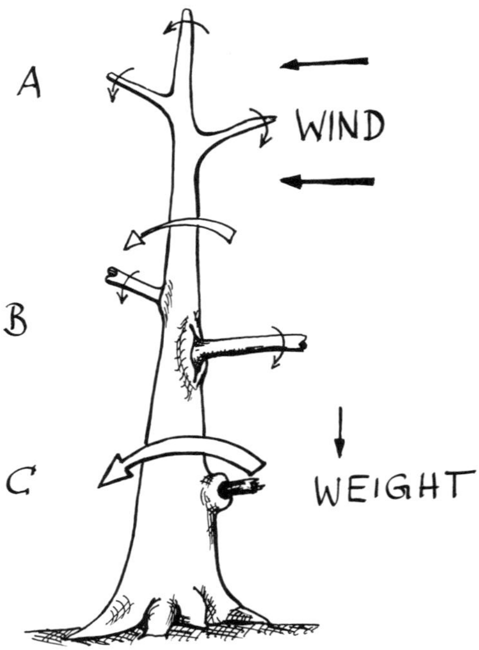

THE SOCIAL DECLINE OF BRANCHES

The branches showing the smoothest, best-optimized joints at the top of the tree (A) compare in length with the leading shoot. Apart from bearing their own weight, they are exposed to similar wind bending moments. Receiving much light above, they essentially contribute to the tree's survival. They are socially important branches which must be maintained through shape-optimized, rounded joints admitting no notch stresses. Branch-stem joints such as these are most unlikely to fail. They ideally suit the requirements of lightweight design. The branches in the middle of the stem are less important with respect to light reception. The bending loads to which they are exposed are not as significant as those acting on the stem at this level (B). It is especially true for dense stands that stems at (B) have long lever arms reaching into the top, but relatively short branches protruding. In other words, the down--stem force flow is stronger than the lateral force flow introduced by the midbranches. In the process of shape optimization at (B), it is the vertical force flow in the stem rather than the minor force flow coming out of the branches which determines the design of the stem-branch joint. In the hierarchy of vital components, the branches are treated as holes, or dead branches, and are bypassed by bulge formation at their bases. The vertical force flow is diverted at both sides of the branches, and notch stresses are reduced by the newly added axial reinforcements. Even less regard is given to the attachment of branches near the butt of a tree. Some of them are already decayed and rotten, and most are largely protected from wind by surrounding bushes in dense stands. The tree grows around them forming a collar which reaches the point where the down-stem force flow enters the branch base. This circumferential notch acts as a predetermined breaking point preparing the breakoff and imminent loss of the branch. The stages of this "natural pruning" are described in more detail in the following two examples.

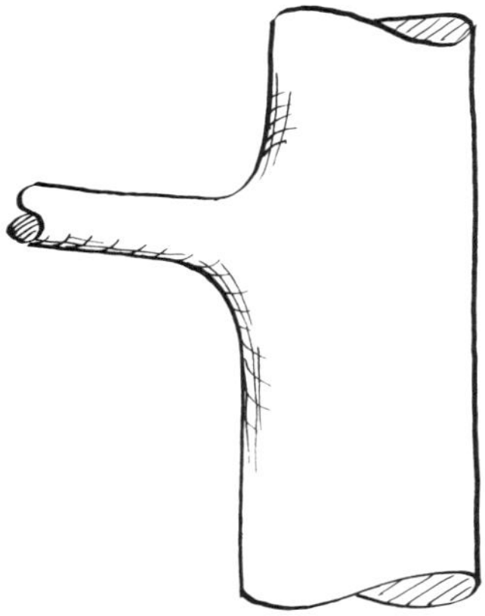

A

ATTACHMENT OF BRANCHES IN THE TREE TOP AND IN THE MIDDLE OF THE STEM

Highly loaded as they are, both the branches located in the top of trees in dense stands (A), and the lower branches of solitary trees extending laterally toward the light like long horizontal lever arms, are perfectly jointed to the stem, showing highly shape-optimized transition contours which do not admit any notch stresses. It is evident that force flows transferred from such excellently optimized branches into the stem are significant and compete with the vertical force flow in the stem. If the down-stem force flow dominates, another branch-stem joint is formed (B), which is found mid-stem in dense stands or at the very butt in the case of solitary trees. This type of joint is in no way less shape-optimized than joint (A). However, it is a design adapted to the weaker force flow in the branch at this level. The joint is characterized by two axial bulges diverting the down-stem force flow which passes by. The branch-to-stem force flow is duly controlled by smooth contours at the upper and lower sides of the branch. A view of these contours is blocked in the sketch by the axially reinforcing bulges approaching each other just above and below the branch.

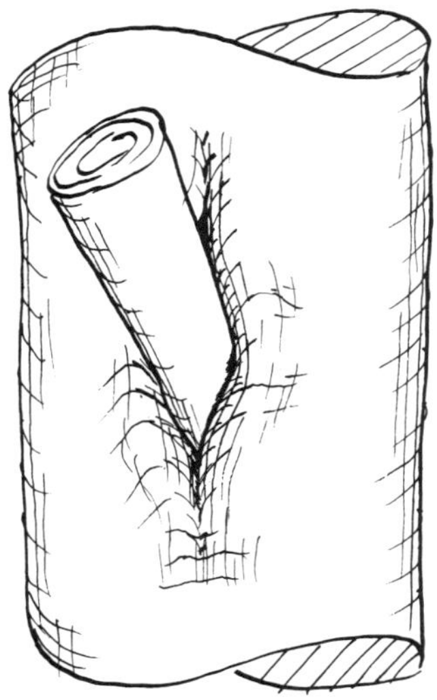

B

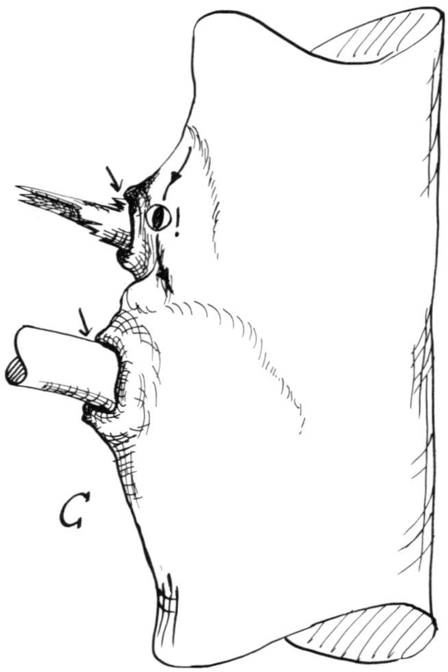

LOWER BRANCHES DISCARDED

Contemplating the tree sketched on the left, we are somehow reminded of the life of man. The lower branches, once growing strongly toward the light, and of service to the tree in many respects are decaying. They have developed into passive good-for-nothing members which are no longer stressed or capable of resisting further. They are surrounded by narrow, sharp-edged collars whose notch stresses act as predetermined breaking points signalling them to break. Moving game, wind, snow, or dropping of other branches will sooner or later make them fail at the very edges of their collars. The collars will immediately close the wound and protect the stem from infectious decay. The old branches will be forgotten, and nothing other than two healed wounds will tell their story. Expecting overall shape optimization and stress reduction to be the rule, one might be surprised at the sharp notches indicated by the arrows. Nevertheless, their presence can be justified by the following experiment: A circular notch (!) punched into the collar develops a wound spindle which points in the direction of the dominant force flow. The spindle shown in the sketch runs parallel to the stem. This indicates that there is hardly any force flow passing through the branch. Since, under wind loading, the notch at the edge of the collar is unstressed, notch stresses fail to occur. Its presence does not violate or interfere with the tree design rules established so far. The unavoidable, final predetermined-breaking-point effect of the collar notch (arrows) has nothing to do with the normal service loads which determine the orientation of the spindle, but is rather caused by the combined effect of external accidental loads and impacts such as touching game, snow, or ice, and related notch stresses. Discarding the old branch, the tree is prevented from decaying, and from wasting energy and material for non-working parts.

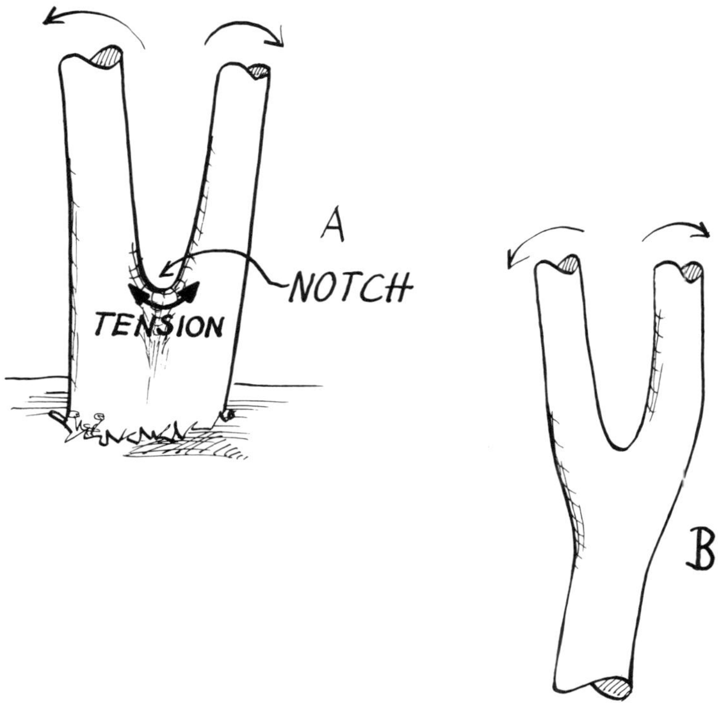

THE TENSILE FORK - A STEM FORK WITHOUT NOTCH STRESSES

Tensile forks shall be defined as stem forks consisting of two or more jointed stems bending away from each other, which leads to tensile stresses acting on the concave part (notch) of the joint. From the design point of view such deeply notched forks appear rather critical. An engineer unfamiliar with the phenomenon of self-optimization would expect significant notch stresses. However, finite element calculations by Mattheck and Vorberg (1990), and application of the CAO method, gave proof of a virtually constant stress state all along the inner contour. It is evident that forks such as these are excellently shape-optimized components with very good fail-safe features. Regarding the completely different outer, but quasi-identical, inner contours of (A) and (B), one concludes that the concave inner contour of the notch largely determines the shape-optimized design of the stem fork, and prevents the fork from breaking. FEM calculations furnish proof that the outer contours are not as important for the safety of the tree as the inner profile.

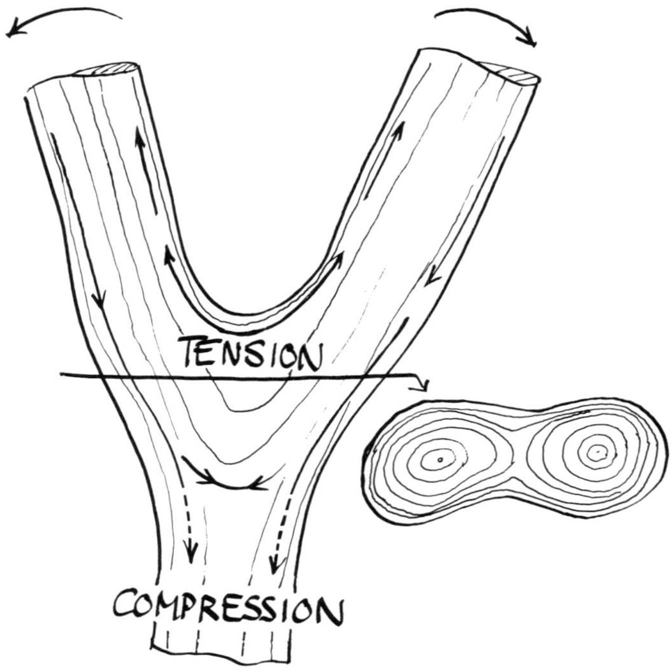

RELATIONSHIP BETWEEN FORCE FLOW AND TREE RING ARRANGEMENT IN TENSILE FORKS

The avoidance of shear stresses between neighbouring tree rings seems to be some kind of a general design rule of axial and circumferential tree ring arrangement along the principal stress trajectories (i.e. the force flow), which are free from shear per definition. In the case of tensile forks, it is the tension-loaded section connecting the stems, which typically consists of tree rings running continuously from one stem into the other. They exactly follow the trajectories of the very tensile stresses characterizing the tensile-fork species. It is just this phenomenon of continuous tree rings running from stem to stem which accounts for the superior quality of the tensile fork. The compression fork discussed in the following lacks this distinctive feature.

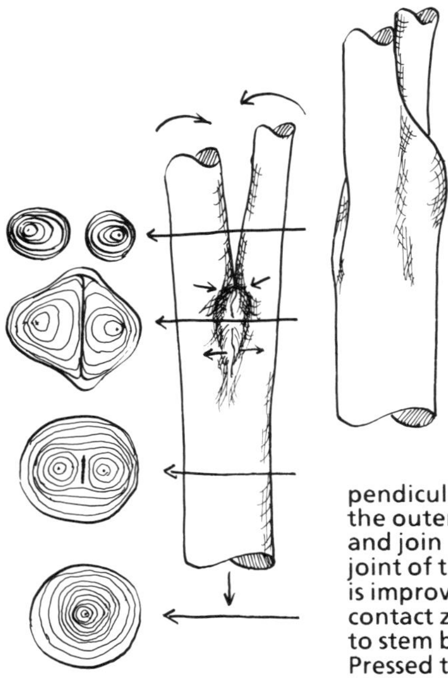

THE COMPRESSION FORK - A NARROW DESIGN

Compression forks are defined here as tree forks whose divided stems are pressed together by the strong formation of reaction wood. The resulting contact stresses are reduced by contact surface enlargement. The once circular tree rings form a nearly plane contact area for the circumferential tree rings to run perpendicularly (since, in the real sense, "tree rings" are hollow cylinders, one should differentiate between circumferential and axial arrangement). It takes a few years for the tree rings to finally run perpendicularly to the contact area. Once they do, the outermost circumferential tree rings weld, and join in a common ring which tightens the joint of the two stems. The strength of the fork is improved, but bark is still enclosed in the contact zone. It carries compression from stem to stem but is unable to transmit tensile force. Pressed together as they are, especially in the upper contact area, the stems need a balancing tensile stress to act on their lower parts (see arrows). If these tensile stresses act where bark is enclosed, and if unfavourable winds occur at the same time, compression forks turn out to be inferior to tensile forks in strength and shape-optimized design. The right of such ill-designed compression forks to exist is in their dense-stand growth (e.g. stands of red beech) which forces divided stems and branches to grow straight in order to ensure sufficient light reception. Tensile forks in dense stands would be equivalent to losses in height and denial of light. They are found in solitary trees, in wider stands, or in the tree tops of dense stands. Since wind loads in dense stands are limited, compression forks normally survive in spite of their less suitable design. However, they may fail to resist stormy weather if stands are thinned by selective felling. With more space available for the branches to spread, they may also break under worst-case, unused-to tensile loads to which they are not adapted. Vertical cracking of the joint along the enclosed bark would be a typical case of compression fork failure. Although improving light reception in dense stands, which admit mechanical insufficiencies, compression forks, in summary, must be considered weak points in the mechanical design, especially when they are loaded by lateral tension.

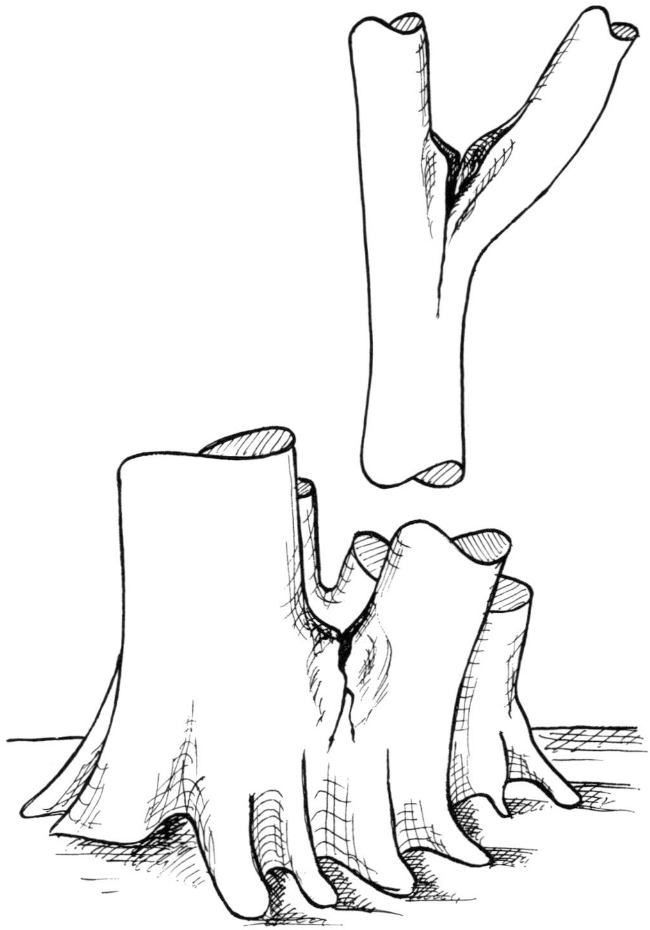

AXIAL CRACKING IN COMPRESSION FORKS

The bark caught between the contacting, divided stems of the fork is often surrounded by a few welded outer tree rings, or fails to be surrounded at all. If, for example, stands are thinned, the divided stems are suddenly free to spread and extend laterally. Tensile loads may then act on the compression fork, which unlike the tensile fork, is not equal to them. Out-of-tune swaying under wind loads can also affect the stems and impose lateral tension. In this case, the divided stems will cleave along the enclosed bark.

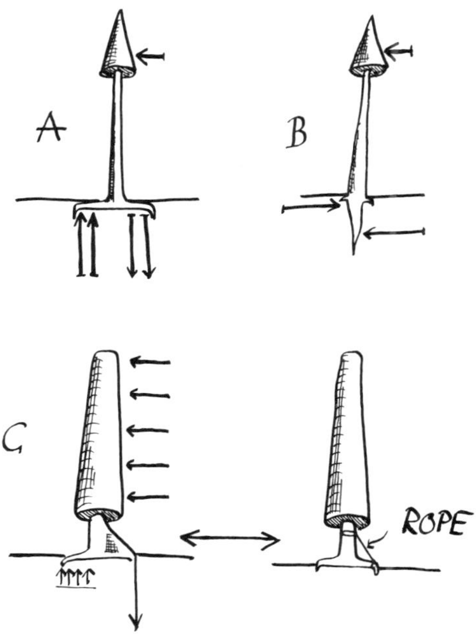

BASIC ROOT TYPES AND RELATED LOADINGS

With wind loads acting on the crowns of trees, stems take the function of long lever arms producing high bending moments which must be counter-balanced by the root-soil moment to prevent trees from falling. Since, as a rule, roots are much shorter than the stem, the counterpoise root forces must be in excess of the force of wind loads. Completely uprooted trees with soil plates attached are often found with flat-rooting trees (A) such as spruce, and, less frequently, in the case of tap-rooted species (B). Buttress roots (C) have the effect of ropes stabilizing a tent pole. With the down-stem force flow passing right into the outermost lateral tips of the root, the flat, horizontal roots are braced like a crane jib, and the full length of the lateral roots is mechanically activated. It is, for example, typical of black poplar to form at least one buttress root to act like a rope or tension cord at the windward side. The conditions of buttress root formation are outlined later in this book. Of course, there are further root designs between extremes (A) and (B). The shape of the root system may vary from tree to tree depending not only on species but also on soil conditions (groundwater level, etc.). The two extreme root stock designs shown here are just to illustrate how wind loads are brought into the ground. In spite of occasional failures, the success of these anchoring systems is a great wonder from the mechanical point of view.

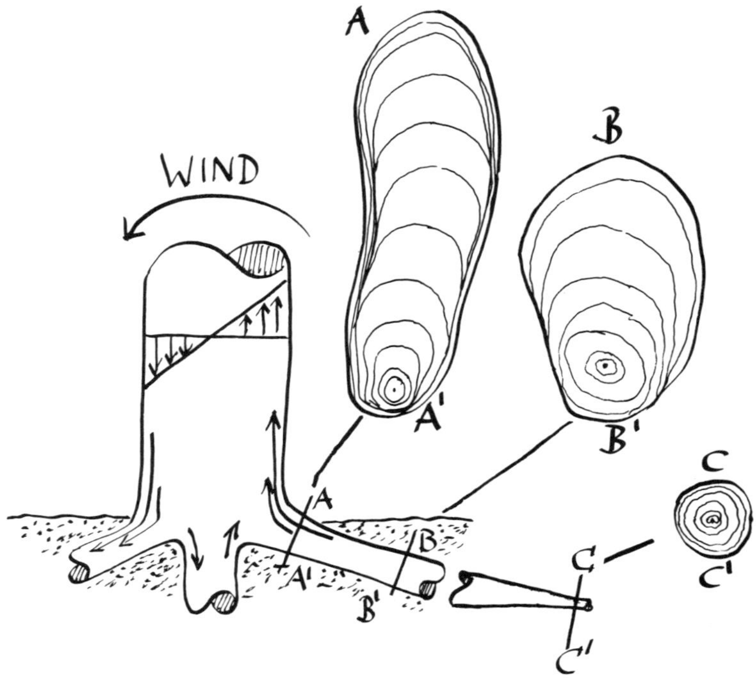

NEAR-STEM ROOT ECCENTRICITY

The wind causes bending stresses in the stem, which act as compressive stresses at the lee side and as tensile stresses at the windward side. They are highest on the stem surface and zero in the middle. The force flow is diverted into the roots (arrows). Entering the shortest way, it will be strongest and will produce the thickest tree rings in their upper parts. The near-stem eccentric cross-section (A-A') thus developing may finally lead to unilateral buttress root formation. Theoretical proof of this was furnished by finite element calculations by Mattheck and Prinz (1991). Farther off, force flow distribution over the root cross-section is more homogeneous, i.e. the tree rings are of quasi-equal circumferential thicknesses. It should be mentioned that combined tensile and bending loads may also lead to eccentric root cross-sections, because superimposed tensile stresses increase the tensile stress of bending while decreasing compressive stress. In accordance with the above, this non-symmetrical stress distribution will also stimulate non-uniform tree ring thicknesses with the tensile, upper side increased. These two explanations somewhat overlap, but the latter can duly account for the preference of buttress roots to form at the windward side where much tension is superimposed on the root bending. On the lee side, the compression force flow enters the ground quite near the stem. Significant vertical ovalization at lee sides can only be expected if there is poor contact between the horizontal root and the ground below.

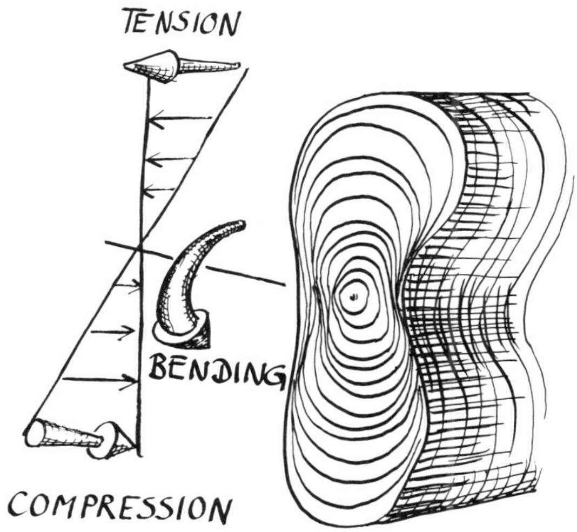

AN I-BEAM-TYPE ROOT CROSS-SECTION
(The author is grateful to Dr. Gardiner, Roslin, and Dr. Wood, Oxford, for encouraging the consideration of these profiles)

Judging by its symmetrical arrangement of tree rings, the cross-section shown in the figure must have belonged to some root far off the stem. The cross-section is characterized by wide top-and-bottom tree rings and near-to-zero thickness in the middle. The model of bending stresses shown on the left (cylindrical arrows) reveals that forces are strongest at the top and bottom and zero at their transitional intersection. This underlying assumption suggests that the cross-section was loaded by bending alone without any axial pull superimposed, and that it was situated far away from the stem. Such 8-shaped roots are frequently found in spruce. Some million years before civil engineers could think of contriving it, nature invented the I-beam and set an example. Finally, it should be mentioned that tree rings in such root cross-sections often reveal changes in loading. A circular shape in the centre, which is followed by eccentric growth indicates an increase in bending after initial loading by tension alone. While these changes may be caused by addition or removal of crown parts or by changes in neighbouring roots, the absolute thickness of individual tree rings is a result of seasonal conditions such as humidity and temperature.

BUTTRESS ROOTS

Acting like natural tension cords tying outermost root tips and stem, these buttress roots brace the lateral roots, and increase their effective length in accordance with the lever-arm principle. Disregarding the fact that wind loads nevertheless act on the stem as the longest lever, roots far from the stem would be too flexible to resist major bending moments without these tension cords. Judging by the amount of buttressing, the tree considered seems to have been exposed to winds blowing stronger from the left than from the right. "Buttress root" is a misleading, if not incorrect term, and is only employed here because it is commonly used. It suggests the presence of compression loads, although the roots would fail by buckling if exposed to them. Buttress roots are designed to act like ropes stabilizing a tent pole. They carry tension, and under tension loading not even blades would buckle, i.e. become mechanically unstable. To conform to international nomenclature, the term buttress root is used, however, but *"tensile root"* would be a better term.

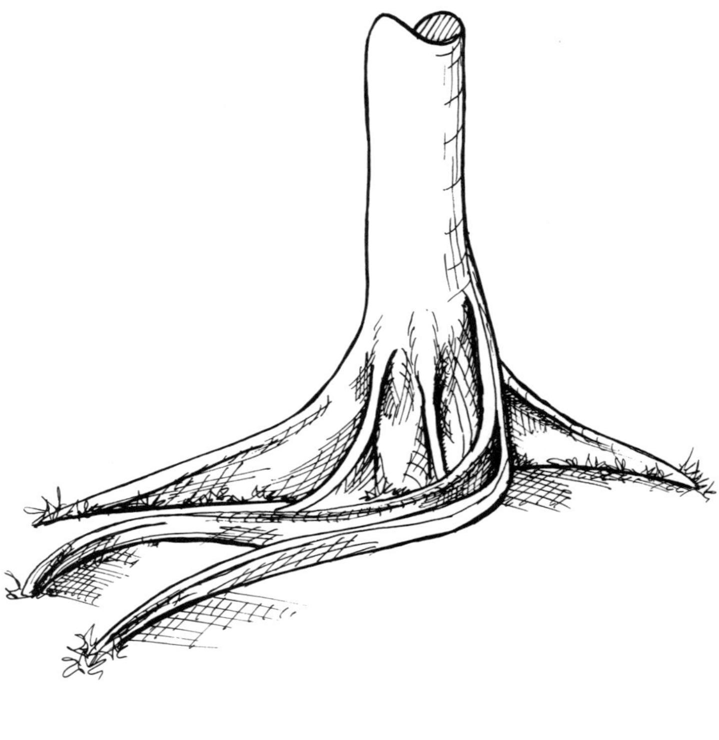

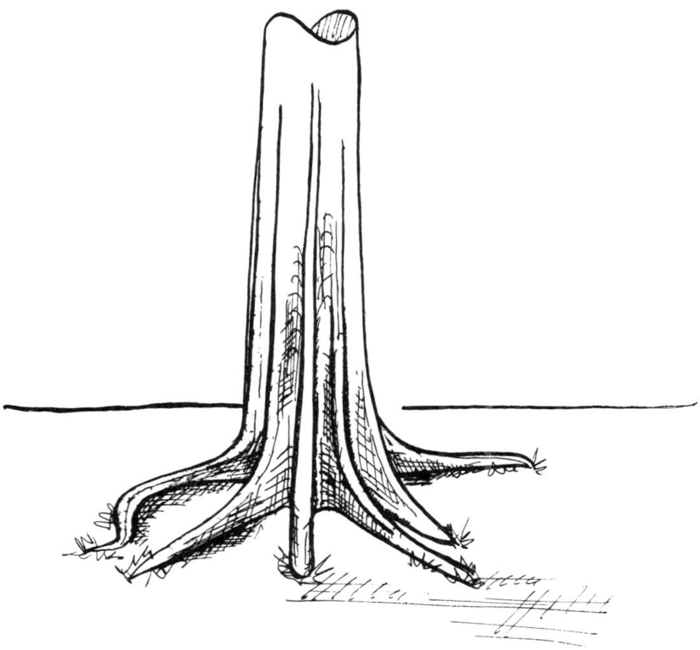

BUTTRESS ROOTS IN SYMMETRIC ARRANGEMENT

This tree was obviously exposed to the wind from all sides to a similar degree. No direction seems to have dominated. Finite element calculations by Mattheck and Prinz (1991) have shown that also in the case of buttress roots, constant stresses are distributed along the upper root edges. Differentiating between compression- and tension-loaded members, buttress roots are advanced, ingenious, lightweight structures, which, although not as sophisticated, can be compared to the mammal skeleton, where muscles and tendons act as tension cords which reduce the bending load which the bones have to carry. Even more than in buttress roots, the principle of the skeleton is manifest in aerial roots. For this quite symmetric arrangement, it should finally be discussed what happens when buttress roots are loaded on the compression side where they do not bear as much weight. The tree shown is well-prepared because it has another buttress root with high tensile stiffness on the opposite side of the butt being appropriately loaded by tension. If, furthermore, the buttress root under compression is in good contact with the ground below, the force flow will probably not run along the full root length but proceed near the stem into the ground. The opposite is true for the tension side, where the tension-loaded buttress root is lifted off the ground.

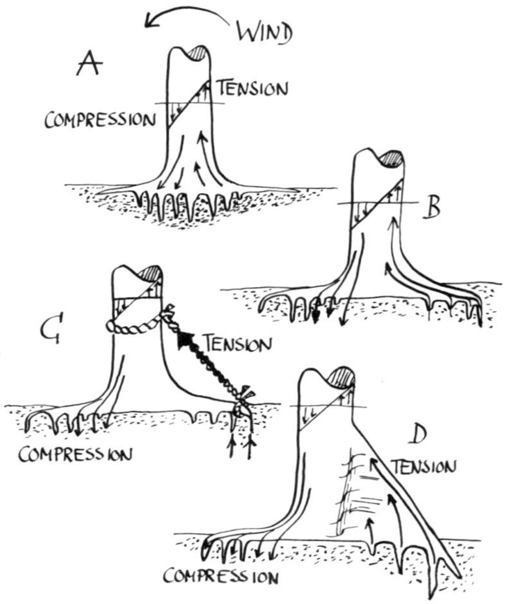

BUTTRESS ROOTS - WHY THEY GROW

In the search for an explanation of the phenomenon of some flat-rooting tree species which form buttress roots while others do not, one comes across two essential mechanical factors. In the case of flat--rooting trees (no tap roots), the main lever arms of the root system point away from the stem instead of entering the soil vertically as a prolongation. Although such structures seem to sufficiently justify the existence of buttress roots, computer calculations at the Karlsruhe Nuclear Research Centre revealed that yet another condition must be fulfilled to stimulate their formation. If the small roots running vertically (sinker roots) are more or less directly below the stem (A), the main force flow runs down along them to enter the soil near the stem axis. Considering that, for lack of mechanical stimulation, the lateral ends of the horizontal roots remain thin and flexible without any major loads to carry, it is evident that the additional condition mentioned above includes the presence of vertical sinker roots which are appropriately arranged at the very lateral ends of the horizontal roots (B). Given this arrangement, the force flow passes the entire length of the horizontal roots, and makes them grow in thickness. A rope (C) connecting root tip and stem would be a suitable mechanical aid in transmitting the tree forces into the soil by using the longest possible horizontal lever arm. This longest lever arm is of course related to the smallest possible root-soil forces necessary to anchor the tree in the ground. Buttress roots formed by dominant upper-side growth (D) are the natural counterparts of this mechanical expedient. They are found in Europe in spruce, beech, and especially in black poplar at the windward or slopeward side of trees, but are rarely seen at the lee side where force flows enter by pressing the horizontal roots against the soil, thus relieving the sinker roots of their anchoring function. Using the CAO method, the above mechanical explanations have been convincingly verified by Mattheck and Prinz (1991).

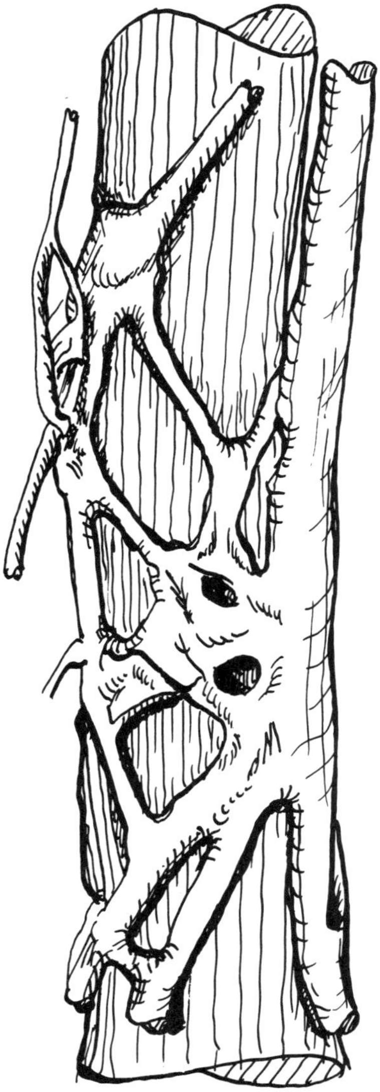

AERIAL ROOTS

Aerial roots are stringy and only a few millimetres in thickness when in the hanging state. Once contacting the soil, they anchor, contract axially, and increase their girth. As shown by Zimmermann et al. (1968), their postanchoring axial contraction is so strong that it can slightly lift a soil-filled container in which they are anchored. It gives them the strength of ropes, which with their larger surface available for adaptive growth can regulate the root system's prestress and anchorage. The figure shows that aerial roots often weld to form frameworks with rounded, shape-optimized joints. This willingness to fuse and weld is of special importance to the strangler fig, which starts life on top of a host trunk from seeds brought by birds. In the course of time, the entwined host tree is strangled, dies and decays. While many of the vertical tensile roots (aerial roots) become increasingly loaded by compression, the danger of buckling is controlled by lateral bridges which are formed by horizontal roots welded to the vertical ones. These are the roots which provide stability after decay of the host tree. From the point of view of statics, such frameworks may well be considered to represent the most sophisticated stage of adaptive shape optimization, nearly comparable to the adaptive potential of trabecular bone.

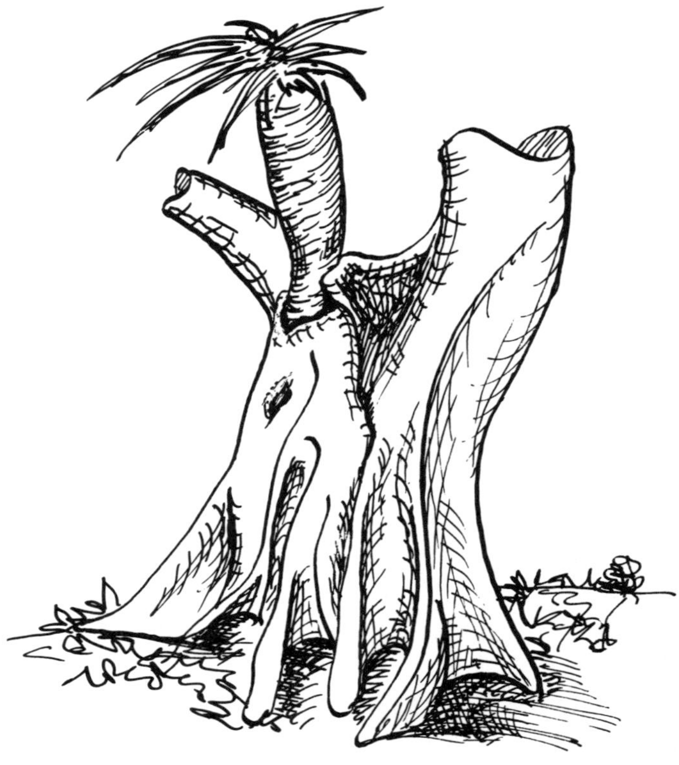

THE INCARCERATED PALM TREE

It cannot clearly be discerned whether the palm tree, which is completely surrounded by the strangler fig was really the host tree or whether it just happened to grow here. Regardless, it is surprising how it managed to survive inside the tight collar formed by the fig tree. One reason could be that the fig thickens laterally but the palm does not. It is doubtful whether another strangled tree capable of secondary lateral thickening would survive in such a critical situation. In spite of the fact that the palm does not increase its girth, the wind load to which its crown is exposed seems to introduce considerable lateral contact stresses into the strangling collar. This is indicated by the smooth contour formed on the right-hand side of the collar. Here, the fig obviously minimizes the lateral contact stresses by increasing the contact zone. Although these two different species, of course, do not fuse, the contact zone is well-adapted to the force flow.

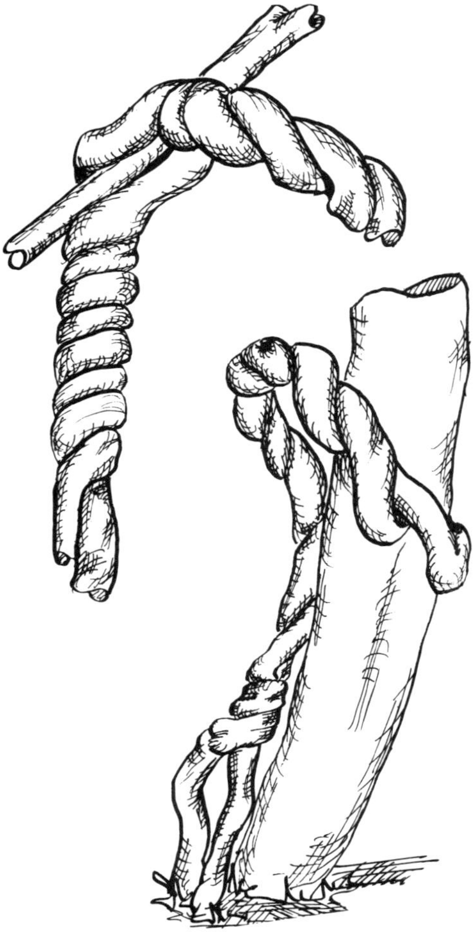

DISLOYAL AERIAL ROOT PAIRS

These intimately twisted pairs of aerial roots seem to be in a phase of unstable adaptive growth. Their coils impede one another by increasing thickness. Especially in the example on the left there is hardly any space between them. Since aerial roots may produce tension wood when under loads, and since any mechanical contact induces faster growth near the contact area, the girth of each root increases more and more while the coil diameter remains quasi-unchanged. This leads to self-strangulation of the root pair, increases the contact stresses between the coils, and may result in the welding of neighbouring coils. Assuming it is of the same species, the slanted branch which got caught in the root pair shown on the left will also be integrated by welding, i.e. its axial transport system will be strangled, and at least its off-stem end will die. Any disproportionately tight bond of two individual aerial roots is of mechanical disadvantage, and hinders the roots from supporting and anchoring the tree. The root pairs shown on the right are similarly twisted around each other, but reduced their mechanical self--impediment by torsional buckling. It is generally known that ropes form sharp kinks if twisted too much. Although this type of mechanical outlet leads to a seemingly useless arcade, it is a solution involving a lesser degree of self-strangulation. In spite of the disloyalty regarding their stabilizing function, the root pairs will keep on fulfilling their medium-transporting function all along the twisted distance.

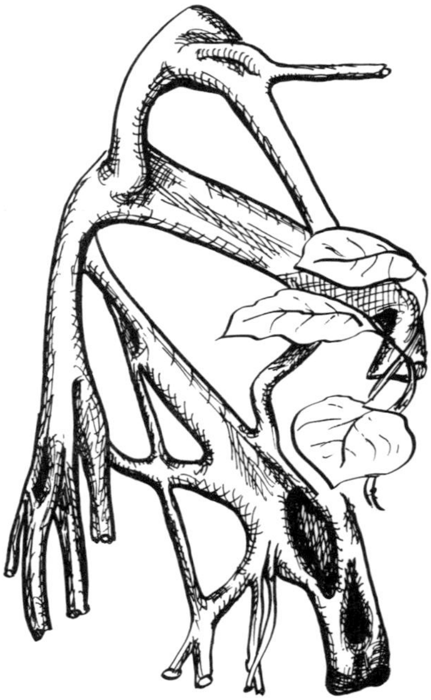

FRAMEWORKS OF AERIAL ROOTS - WELDED COORDINATE SINGLES

Unlike the average tree, which only has a relatively small area of surface cambium capable of adaptive growth, trees anchored by aerial roots have a much larger adaptive potential due to the increased surface formed by the network of contractile strings, ropes, and trusses. In addition, aerial roots readily weld at points of contact, which leads to high-tech frameworks with smooth, shape-optimized joints without any notch stresses. Engineers and designers cannot but marvel at and admire such refined biological structures, which seem to eclipse any of their own awkward, oversized maximum-safety designs. However, the mechanism of adaptive growth has been successfully copied by computer simulation at the Karlsruhe Nuclear Research Centre. Compared with non-optimized structures, engineering components shape-optimized in this way have shown a 40-fold longer endurance in fatigue tests. Some of the specimens tested turned out to be quasi-indestructible. The new method, named CAO (Computer-Aided Optimization) has been well accepted by the industry. All of the renowned companies using it have benefited from the design rules taught by the body language of trees. The figure below shows a knot obviously made for fun into one of the still non-anchored hanging strings of thin aerial roots. After anchoring in the ground the root was stressed and started thickening. The knot was tightened so much that parts of it in fact even weld. In some years, however, probably no more than a localized thickening will remain as witness.

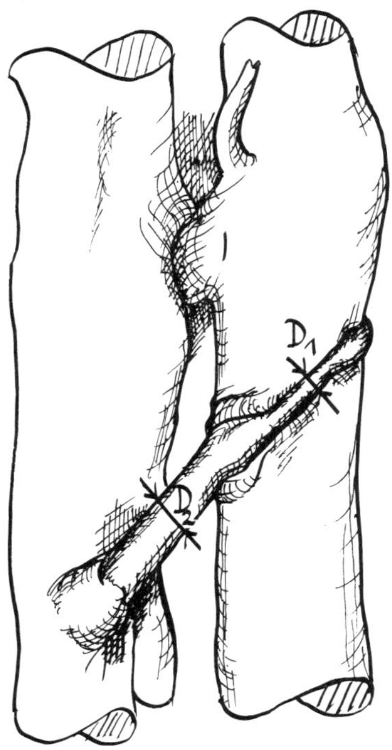

A ROOT GROWING WITH ITS TASKS

Two thick vertical aerial roots are interconnected by a smaller slanted root which is tightly attached to the right vertical root through frictional contact or partial welding. Thus supported and protected, the right part of the slanted root remains nearly unstressed, and has a correspondingly small diameter D_1. In the gap between, the slanted root has to carry the load alone, meaning that circumferential growth is stimulated to create a larger diameter D_2. $D_1 < D_2$ shows the expediency of diametrical changes, and proves that the proverbial growing-with-one's-tasks phenomenon is applicable to roots as well. This small part of a tree is a model example of the constant stress hypothesis as a design rule for biological load carriers. Obviously, the study of tree design is excellent practice for engineers dealing with structural mechanics. Contemplating thicker (higher loaded) or thinner (less loaded) members of the tree will give a good idea of force flows in mechanical structures. Although the same understanding of force flow and design relationships can be gained from the results of numerical stress analyses, for example, using the finite element method, trees in their natural surroundings are the more effective "appliances for teaching". They are an enjoyable adventure and an impressive experience in one.

GROWTH IN COMPETITION WITH DECAY

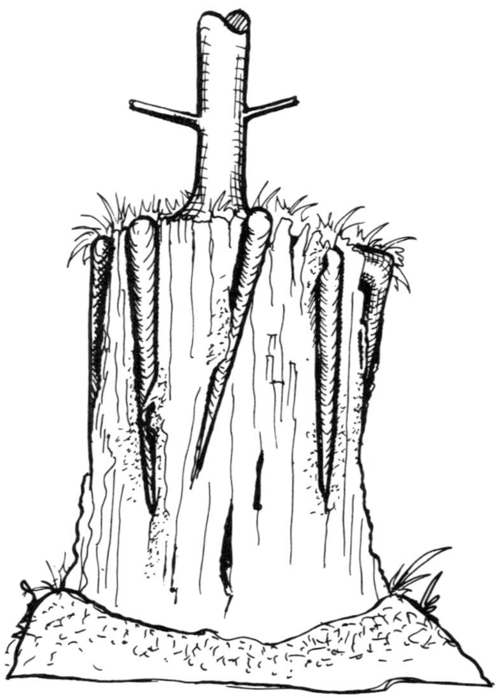

Aerial roots can be either programmed genetically as in the previous example, or can be developed by any other tree species as similar so-called stilt roots. The phenomenon of stilt root formation was observed by the author in the North American rain forests of Queen Charlotte Island: Seeds which had fallen on an old tree stump started sprouting. The lateral roots of the developing shoot grow radially towards the edge of the stump. Well--anchored, they proceed bending to grow down the old stump. Competition then begins between the increase in the length of the roots and the decay of their host stump. If the lowest root tips succeed in reaching the soil below the stump and in anchoring the tree in the soil, and if the still very thin root system is quick to stiffen by thickness growth, the mechanical support of the host stump will no longer be necessary, and the young tree will have a good chance to survive. If the roots are too slow to reach the soil in time, or if the root system is too weak to resist storm or snow loads, the tree may be heavily damaged or broken unless brought upright again by the formation of reaction wood (negative geotropism). Another type of stilt-rooted tree (not shown) can be produced by erosion of the soil below trees already anchored in the ground. Especially trees growing on mountain slopes are predisposed to becoming the victims of erosive action. A stone placed below or near the tree can be a successful prophylactic measure which prevents or at least delays the washout of the root system.

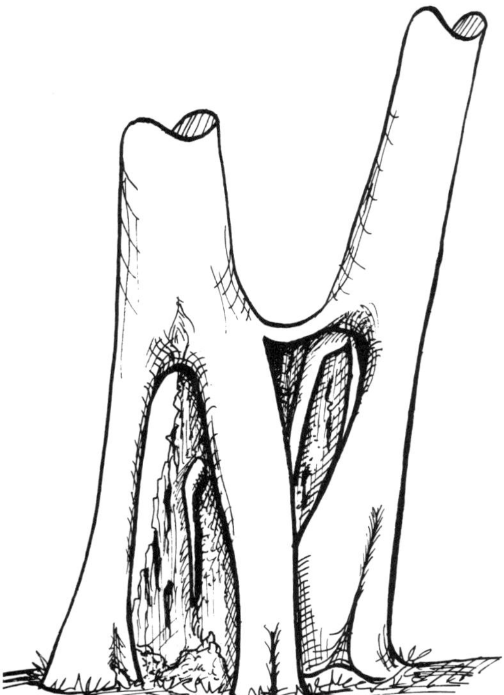

STILT ROOTS

This conifer, once sprouting on an old tree stump, competed successfully with stump decay. It reached the soil with its root tips, and anchored there. Its particularly well-developed outermost roots reveal once again that stresses are highest in the outermost surface regions of components under bending loads. While the outermost roots responded to the high-stress stimulus by an increase in thickness, the roots in the middle of the old stump did not gain in strength since they were quasi-unstressed near the neutral axis of bending. In addition, they are still embedded in the remnants of the decaying host, which contributes its share to protect them. In accordance with the constant stress hypothesis, the diameters of the thinner inner and thicker outer roots are such that their stresses are equal on a time average. It is interesting that, in trying to reach this state, overground roots develop like tree stems (bark formation, etc.) while underground roots look like real roots. Trees of this type are called stilt-rooted trees. Of course, the stilt roots shown here were formed because biomechanical factors predestined the tree to start its life on an old stump. Other trees are stilt-rooted by nature, for example some mangroves which adapt to wet surroundings in this way.

INSUFFICIENT STAYING POWER OF A YOUNG TREE

In this case, the shoot on its old host stump was so close to another adult tree that it was partially protected by it from mechanical loading by frictional contact. Lack of mechanical stimulation hindered the development of a strong root system of its own, thus it relied on its neighbour for support. When the host stump decayed and was finally washed down the steep slope on which it stood, the roots of the shoot, reaching into the ground, could have supplied the tree with nutrients, but they could not grow fast enough to compensate the loss of support by the old stump. In its struggle for mechanical survival the young, weak tree started growing around the neighbouring adult tree which had supported it from the very beginning. Although this biological consolidation saved the life of the weak tree, it increased its mechanical dependence on the larger protector and prevented it from maturing in a mechanical sense. Human society would call this a "pampered baby of the family". At some points, the young and the older tree seem just to start welding, i.e. to form common cambium and produce common tree rings. With fates fused this way, the dependent tree will probably die early together with its much older protector.

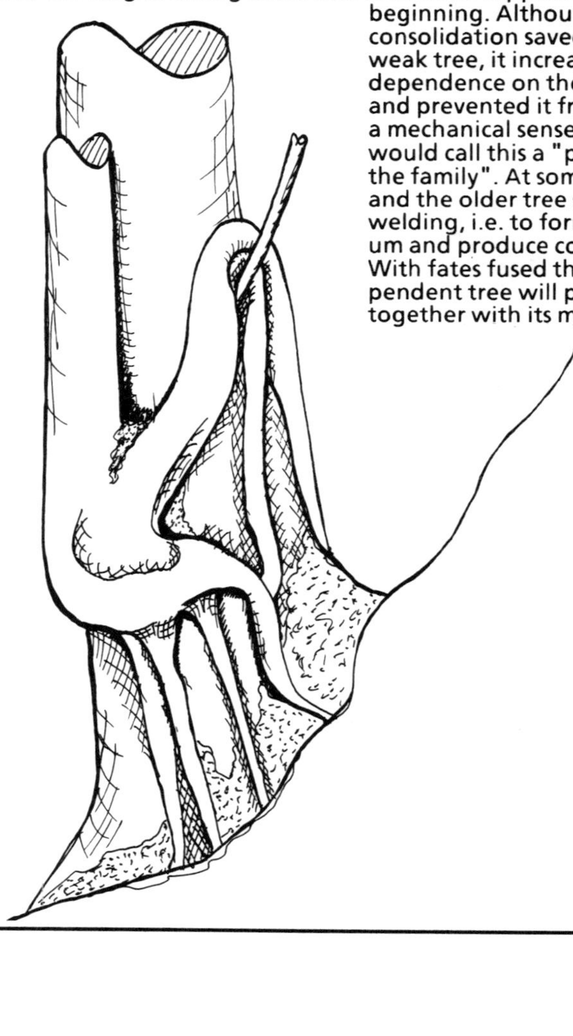

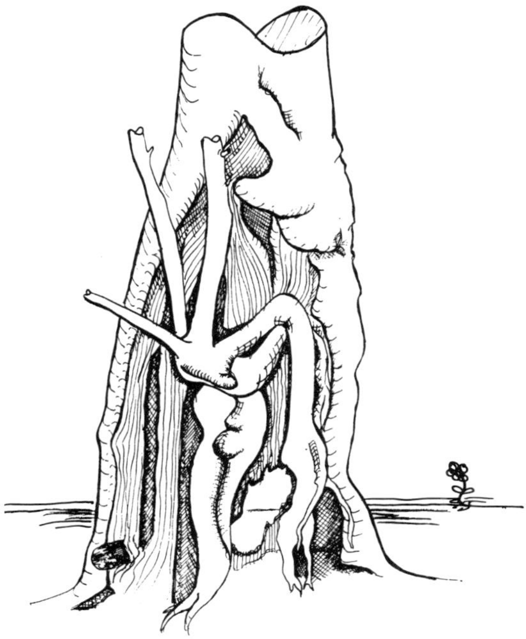

A HOLLOW STEM AS THE PLACE OF ORIGIN

Instead of falling on firm ground, this tree started growing in the decaying hollow stem of an old beech standing on a highland pasture in southern Germany. Pastural beeches are knotty trees, which often consist of many single stems welded together as the combined result of repeated damage caused by feeding highland cattle and subsequent restorative growth, as reported by Schwabe and Kratochwil (1987). The more the bottom of the cavity decayed, the more the roots of the shoot inside approached the ground to anchor there and grow stronger in the remaining half of the old stem. Although still alive, the adult stem of course has no adaptive cambium on the inside of the hollow stem with which the young beech can weld as in the previous example with external contact. Although it appears to be safely surrounded and weather-protected by the remnants of the old tree, it is on its own in the biological sense. Unless it becomes buried under the ruins of the old tree, the young beech may survive the death of the hollow stem due to its mechanical strength. In all these stilt-rooted trees, it is the task of adaptive growth to increase the root system's stiffness the degree to which the support by the host stump decreases due to rot, erosion, or infestation by insects. Stilt roots cannot exist without their anchorage in the ground. However, they also have to be stiff enough above ground in order to stabilize the tree without any further external support.

ROOT EDGE FORMATION ON THE SHORE

This conifer is growing right on the shore of a saltwater bay of Queen Charlotte Island. Quite obviously, its roots avoid the tidal rise and fall of the salty water (arrows) by forming a sharp, partially welded edge along the area swept by it. The ends of the roots point away from the water towards the shore. Such well-defined root edges are not only typical of trees growing on the shore of this island, but were also found by the author on dry riverbanks in southwest Africa, as well as along narrow passes in the Palatinate Forest, where heavy rains are likely to cause soil erosion, leaving the roots without soil, water, and, moreover, without the mechanical stimulation encouraging their growth. Roots right outside the erosive zones of shores, riverbanks, or paths are exposed to stresses which are even higher than those weighing on roots farther off. Their additional strength is due to unfavourable lever arms, and the extra loads imposed upon them by the inactivated rootless erosive zone. This relation between load and lack of stimulation explains the formation of sophisticated, high-strength edges characterizing the bordering trees.

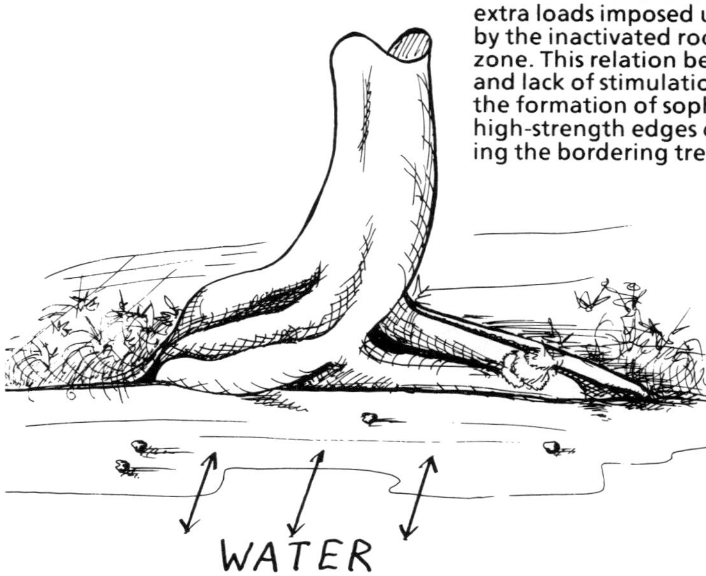

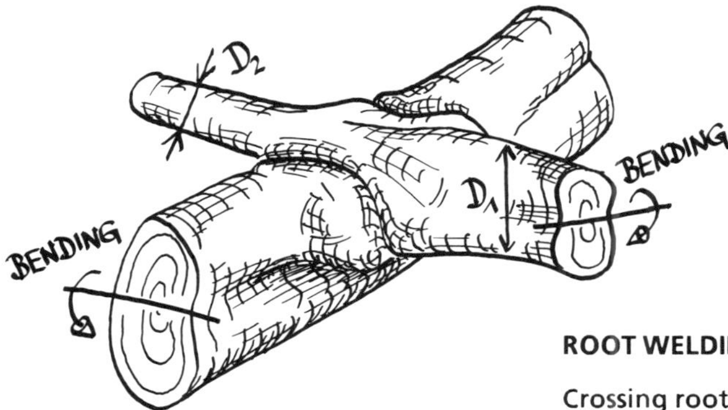

ROOT WELDINGS

Crossing roots often weld, e.g. these beech roots. The hatched area in the cross-section denotes the dimensions that its crossing members must have had when first contacting. Unlike roots of spruce, beech roots of this type hardly show any tree rings. Such weldings develop by passing through the following stages: mechanical contact, wear-induced wound, enlargement of the contact area by local increase in tree ring thickness and smoothing of all sharp-edged points and notches, joint-type form fit and formation of common tree rings, and complete fusion into a shape-optimized design free from notch stresses. These beech root joints are a very illustrative example of root welding and its different stages. In addition, the change in diameter $D_1 > D_2$ in the upper example indicates that the load experienced by the roots may be extremely different on both sides of the joint. The ovalization of the thicker root cross-sections confirms the intense bending loads to which the roots are exposed.

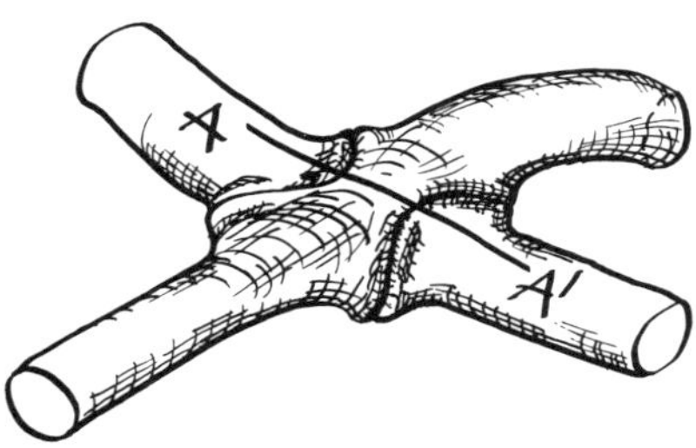

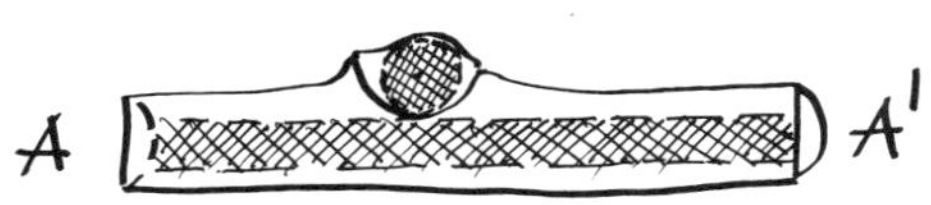

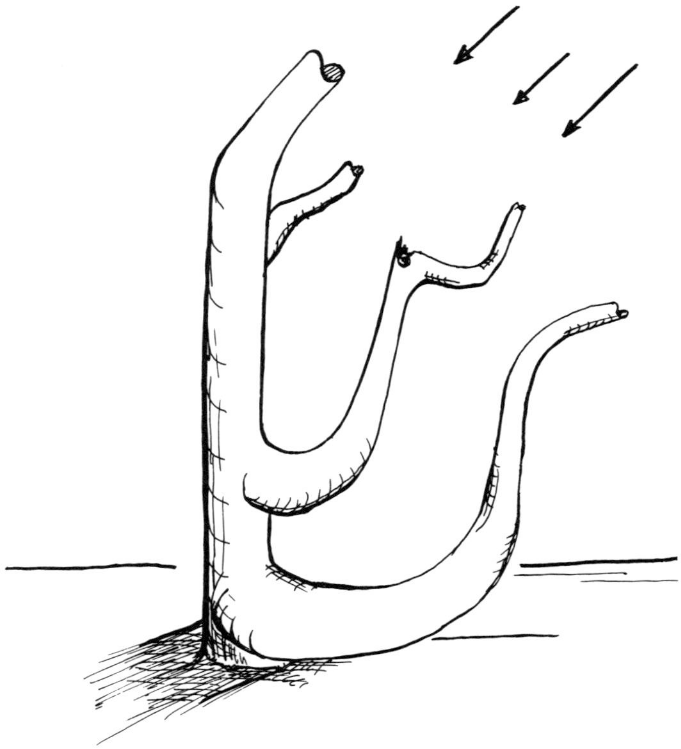

PHOTOTROPIC ARRANGEMENT OF BRANCHES

With respect to the criteria of negative geotropism, this pine is a truly faulty design, a complete crown-over-root-stock failure. Growing at the edge of a dense stand, the tree was induced by phototropism to extend its branches towards the light. Ignoring all bending moments, they assumed a well-tuned and quasi-aligned receiving position throughout. Once growing perpendicularly to the plane of projection, the lowest branch made a 90° turn to the sunny side, and compensated all additional bending and torsion by extra circumferential growth. Compared with other growth-regulating mechanisms, phototropism obviously plays a more dominant role. This is not surprising if one considers that even the best mechanical design is useless without sufficient light reception and photosynthesis. Even the best-optimized parts of a crown are a useless burden when forced to grow in the shade.

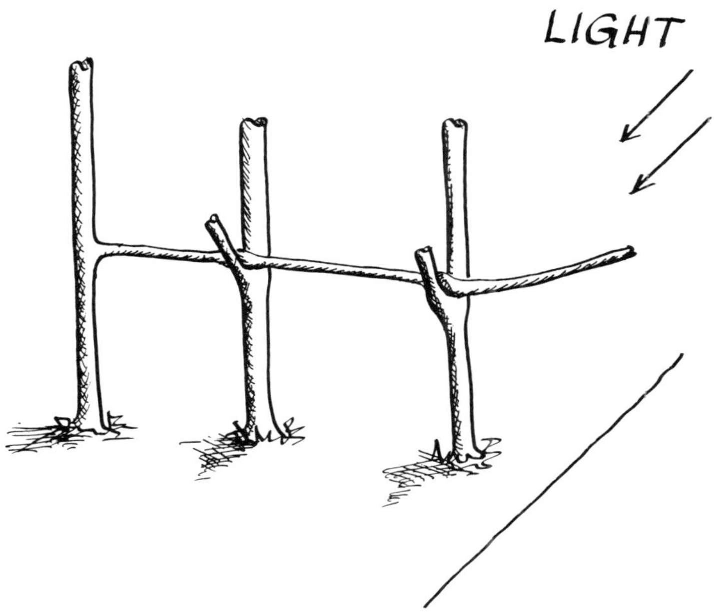

A PIPELINE LEADING TOWARDS THE LIGHT

The left-hand stem is standing in a hornbeam forest about 4 m away from the road on the right. Accidentally, one of its branches pointing south became entangled in the tree fork of a neighbouring hornbeam. This support obviously encouraged the branch to extend in length and grow towards the sunny roadside instead of increasing in thickness. Growing in that direction, it reaches another tree still closer to the roadside, which also supports it. Some further axial growth will now suffice for the end of the branch to be fully exposed to the light. The only difference between the supported branch and real roadside branches is the longer transporting distance from the stem. Wind-induced wear and contact stresses locally increased the cross-sections of the supported branch at its resting points. In spite of the fact that the resting points, too, are bulging to integrate the branch, no real welding takes place. As frequently observed by the author in similar cases, welding may cut off at least the end of the "pipeline" by interrupting its transport system. The receiving part nearest the light would probably die if common tree rings were formed. A lightweight framework or frame design would be the only - mechanical - advantage of a fully welded construction. The increased length, here in the horizontal direction, has also been observed by the author in trees attached to stakes. They normally grow higher than neighbouring trees, which have to resist the wind loads alone, without supporting stakes. It seems that material saved by reduced thickness growth is used by the tree to increase its height.

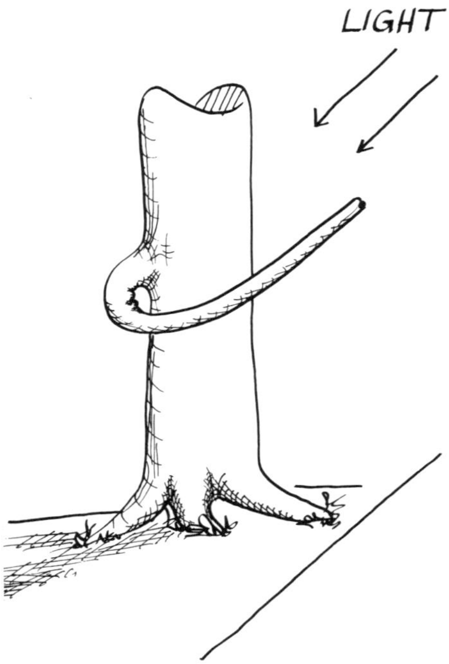

PHOTOTROPIC BENDING OF BRANCHES

This red beech is in the bright sun of a nearby forest path. Its front side faces the path, while its rear is in the shade of the dense, dark stand to which it belongs. The branch protruding on the shady side makes a nearly 180° phototropic bend past the stem towards the light. The extra bending and twisting loads to which it is now exposed due to stronger winds in the open forest path, are compensated by circumferential growth, which, being the only chance for the branch to contribute its share to light reception, can be justified from an energy-balance point of view. It is the dominance of phototropism which accounts for the bizarre designs of some trees, and the mechanical disadvantages for which they must compensate for the sake of light reception. Trees responding to phototropism in spite of competitive drawbacks due to additional supporting material and energy give proof of the superiority of light reception over mechanical advantages. Branches with such sharp bends are exposed to the danger of axial splitting when straightened by snow or wind loading. The need for light reception seems to ignore even the possibility of mechanical destruction. Such branches are *hazard beams* in the real sense of the meaning. They will be referred to as such and introduced in detail later in this book.

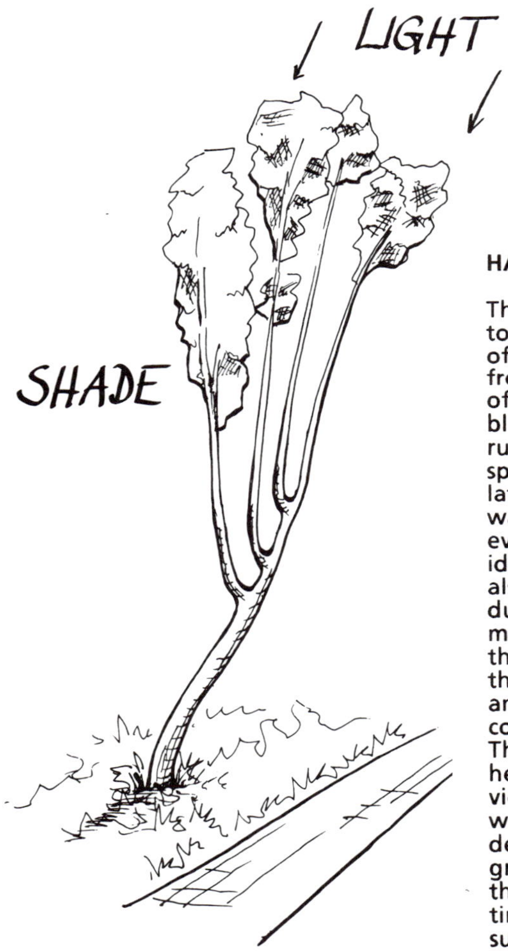

HARP TREES

This roadside birch belongs to a dark and dense stand of tall trees. Struggling to free itself from the shadows of its large neighbours which block the light, it ignored the rules of geotropism, thus responding to phototropic stimulation, and directing itself towards the road. Although its ever-growing, tall stem proved ideal for light absorption, it also increased its weight, producing a long lever arm which, measured from the butt to the centre of gravity, bent the stem downward more and more. Here, geotropism could no longer be suppressed. The long lever arm and the heavy weight of the crown violate the rules of geotropism, which for minimum bending demand that the centre of gravity be over the centre of the root stock. Since too much time and energy would be consumed in bending the thick stem backwards, geotropism acts on the branches, arranging them in a manner to counterbalance the bending moments of the stem, and shifting the centre of gravity. At the same time, the branches involved keep growing towards the light. Such trees are called "harp trees" because of their nearly vertical harp-string arrangement of branches. They resemble someone standing on the edge of a cliff looking down but keeping balance by extending his arms backwards. Harp trees may also be formed when a stem has fallen but remains partially rooted in the soil. When the stem is unable to erect itself by reaction wood formation, the branches grow vertically like the strings of a harp.

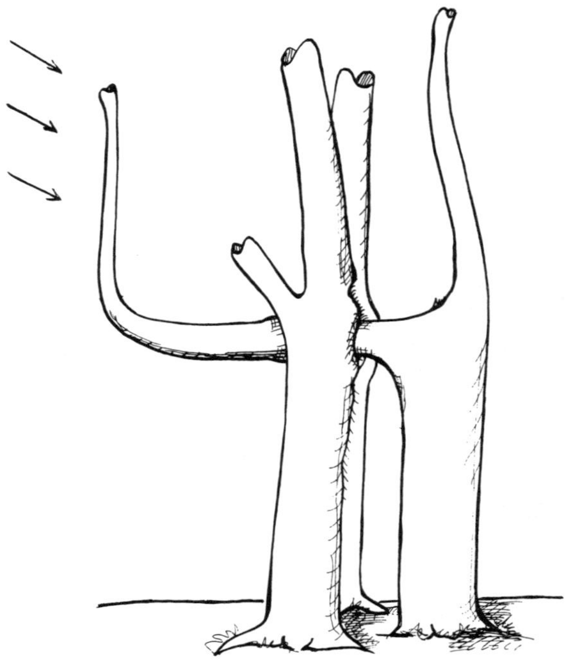

A LONG ARM EXTENDED TOWARDS THE LIGHT

The conifer on the right lost its leading shoot, and remained with two equal side branches suited to compete for leadership. Evidently, the two beeches standing nearby prevented the usual process, and stimulated another development. Unimpeded by the left, the right branch was conceded the thankless task of weak leadership in the shade of the two large trees. The left branch did not try to bring itself upright but extended south, and passed between the two beeches towards the light. As soon as it was out of the shade of the beeches it grew upright. Phototropism then yielded to geotropism, which means that if light reception is sufficient, tree mechanisms concentrate on a good mechanical design. The high bending and torsional moments caused by the length of the horizontal branch are compensated by circumferential growth. This increase in thickness is limited by the width of the narrow gap between the two beeches, which soon make contact with the extended branch. Thus supported, the branch grows thicker on the left than on the right side of its rest. Since the contacting trees are of two different species, welding is avoided, and the transport system remains intact, keeping the left part of the extended branch alive, which would be otherwise strangled.

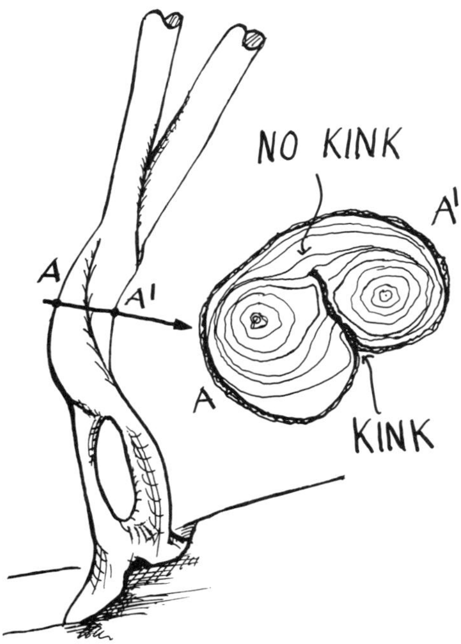

TWISTING AND WELDING

The drawing shows two beech stems which are at acute angles to one another and contact by crossing. Wind-induced wear, form fit, and wound healing lead to their partial welding and subsequent suppression of micromovements of the two stems relative to each other. The interfacial shear stresses and the inclination to bend, resulting from this suppression of movement, resemble the processes known to be characteristic of a bimetallic strip. The bending observed is mainly achieved by formation of reaction wood or, more specifically, tension wood. Its presence explains the phenomenon of slanted-around-upright--stem coiling, which suggests that the slanted stem is continuously stimulated by geotropism to change its orientation but never succeeds because the unyielding, ever-vertical stem again and again prevents it from reaching its ideal position. Owing to their mutual, stress-reducing framework support, the two divided stems do not increase significantly in thickness below the fused region. They are thinner than the free ends above the welded zone. The saw cut shows that the first common tree ring is not formed before the individual tree rings are perpendicular to the interface. It is obvious that tree rings fused into a kinked shape would be exposed to high bending stresses at the kinked zone due to the circumferential compressive growth stresses acting on the surface of trees. In the case under review, shear stresses between the fused tree rings are avoided, and circumferential force flows are free to pass from one stem to the other through a uniform common tree ring, which unlike some non-uniform, fused kink is able to resist circumferential stresses, and minimize shear and bending loads within the tree ring.

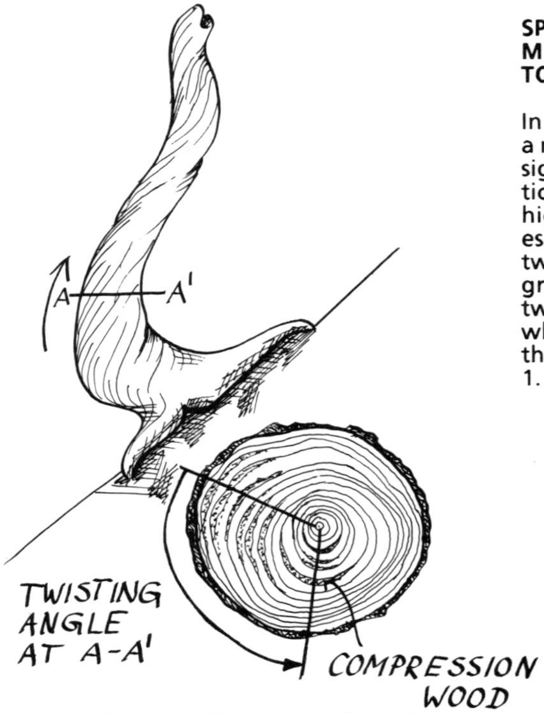

SPIRAL GRAIN - A RELIEF MECHANISM IN RESPONSE TO BENDING STRESSES

In pines, beeches, and in a number of other species, significant geotropic corrections and correspondingly high internal bending stresses are often accompanied by twisted growth, i.e. spiral grain of the stem. There are two mechanical explanations which basically account for this mechanism:

1. Non-circular beams under bending loads tend to twist or tilt out of plane to increase their flexibility. Beams with a rectangular cross-section, which tip over if bent in the direction of their highest stiffness, show the process of combined bending and twisting to release mechanical energy. Geotropic bending is achieved by the formation of reaction wood at one side, i.e. at the upper side of the stem in the case of angiosperms (tension wood), and at the lower side in the case of gymnosperms (compression wood). This unilateral attachment produces an oval cross-section, and causes the tree to bend where it is stiffest. Twisting by 90°, the tree becomes more flexible, because bending stresses are then highest at the flat side of the oval cross-section. New reaction wood, increasing its bending stiffness again, will be followed by another twist relieving the stress, and so on. This bend-and-twist process can be traced by regarding the tree rings of a twisted pine. The cross-section indicates that compression wood (dotted sickle-shaped areas) was always formed at the bottom side of the slanted pine stem. The angle reveals the degree of twisting (A - A') over a number of years.
2. Geotropic correction processes cause significant differences in the axial length of the concave and convex sides of the stem, respectively. The high shear stresses developing between the individual tree rings represent the most critical load case to which wood can be subjected. Twisted growth reduces these shear stresses by minimizing the differences in axial length. The longer part of the tree is bent around the shorter part until the ends are closer. Spiral grain can be considered a shear stress outlet, which, of course, would also contribute to relieving stresses in the case discussed above.

Both explanations overlap to some extent. They may even be two approaches to the same thing, and admittedly are rather speculative. Finite element calculations will be made to verify these assumptions.

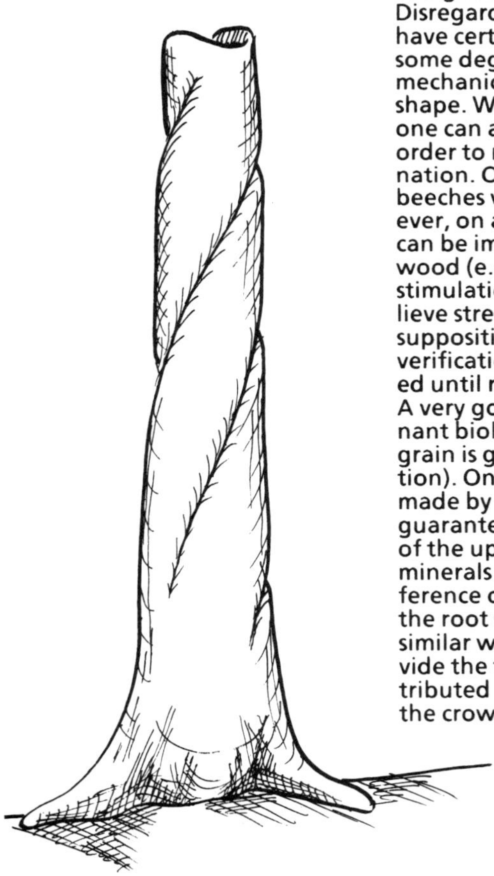

SPIRAL GRAIN OF UPRIGHT TREES

While trees in the process of geotropic upright bending twist in response to bending and shear stresses, it is hard to believe that a similar effect took place in the case of the vertical, straight beech shown on the left. Disregarding genetic factors, which have certainly influenced its coiling to some degree, there must be another, mechanical reason for its bizarre shape. With twisted ropes in mind, one can assume that trees twist in order to reduce the risk of delamination. Of course, there are upright beeches which are not twisted. However, on a smaller scale even these can be imagined to form reaction wood (e.g. in response to phototropic stimulation), ovalize, and twist to relieve stresses. These very conjectural suppositions await their theoretical verification, and should not be adopted until reliable results are available. A very good overview of the predominant biological advantages of spiral grain is given by Kübler (in preparation). One important statement made by Kübler is that spiral grain guarantees a more even distribut on of the upward flow of water and the minerals dissolved over the circumference of the stem even if parts of the root system are ineffective. In a similar way, spiral grain helps to provide the tree with more evenly distributed assimilates even if parts of the crown should fail.

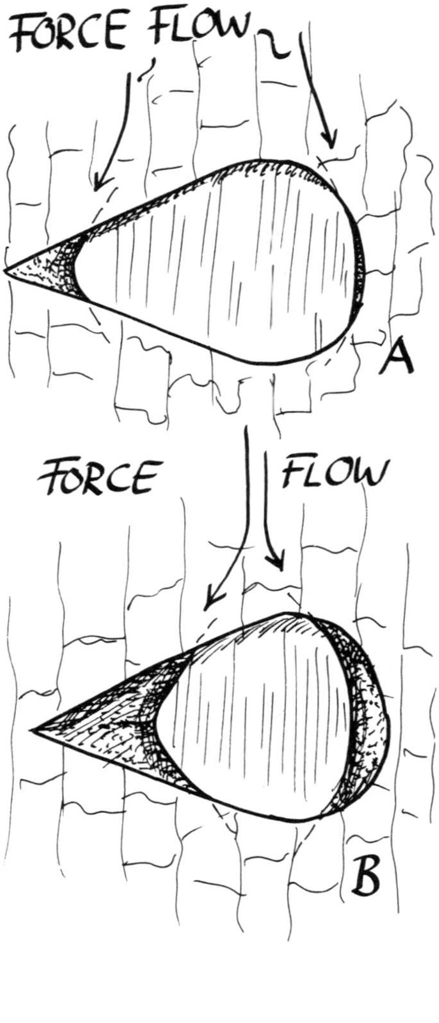

MECHANICAL CONTROL OF WOUND HEALING - THE FISH-SHAPED NOTCH

It has been shown by Mattheck and Korseska (1989) and Mattheck et al. (1990) as well as in previous chapters of this book that circular surface wounds heal into vertical spindles because of high notch stresses on both their sides. Critical discussions with biologists arguing that spindle formation could have purely biological reasons encouraged the author to make an experiment which corroborates his hypothesis of mechanically stimulated wound healing. Non-circular notches were cut into stems. These fish-shaped notches had their highest stresses in their left, pointed corners, and their second highest in the middle of the opposite, circular areas. Wound healing proceeded correspondingly, i.e. wound callus to smooth the notch was first formed in the left corner (A), and afterwards in the middle of the opposite area. Throughout the experiment, wound healing by callus formation was found to be faster in the highly stressed left corner than in the area opposite it (B). Theoretical proof has been furnished by computer simulations carried out at the Karlsruhe Nuclear Research Centre by Mattheck et al. (1990). Since this non-symmetrical progress of wound healing is in excellent agreement with the non-symmetric distribution of notch stresses, there should be no further doubt about regulation of wound healing by mechanical stresses.

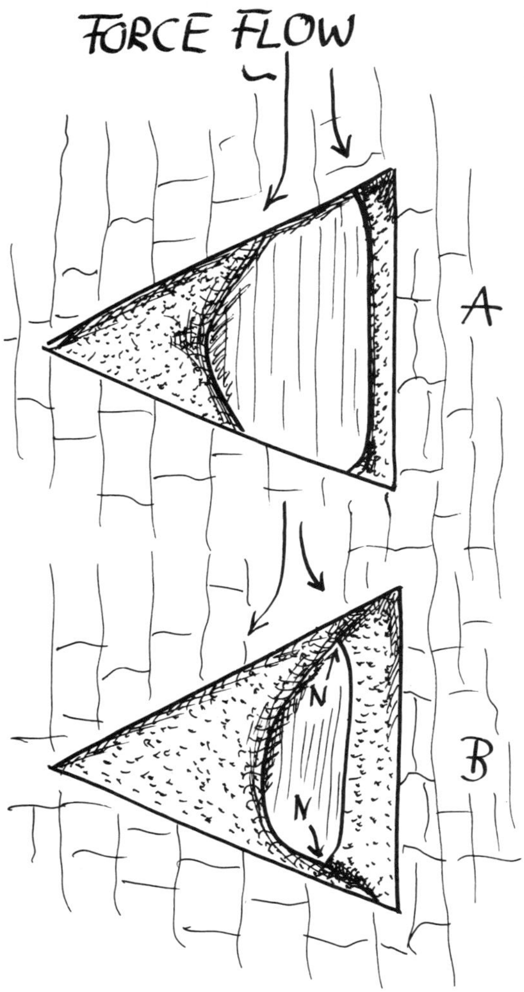

STRESS-CONTROLLED HEALING OF A TRIANGULAR WOUND

A triangular notch cut into the stem of a tree provided further proof of stress-controlled wound healing. As in the case of the fish-shaped notch, stresses are highest in the left corner, which sharply diverts the force flow. The two remaining corners are axially quasi-aligned relative to the force flow. They do not divert it as sharply, and notch stresses are less critical than in the left corner. As expected, wound healing is fastest on the left side, and slower in the top and bottom corners on the right. As wound healing progresses, the callus coming from the left and right sides contacts, and forms a squeezed area with a sharp notch (N). This is surprising at first, because shape optimization is associated with smooth notches rather than pointed tips. However, the tips of these sharp notches point exactly in the direction of the vertical force flow, and can therefore divert it smoothly (cf. the bow of a ship). Notches would certainly have formed differently in a horizontal force flow.
The healing of this triangular notch once again confirms that wound healing is the result of mechanical stimulation. Highest stress peaks are reduced fastest since they are the weakest points in tree design, and the most critical violation of the constant stress rule.

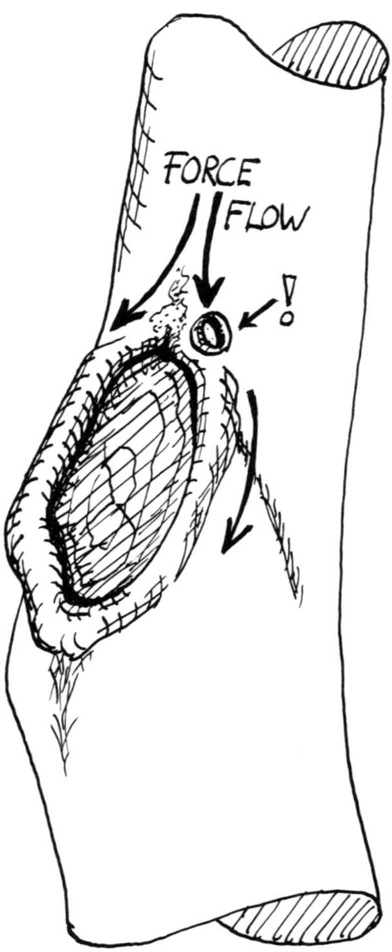

WOUND SPINDLES ACTING AS FORCE FLOW COMPASSES

This drawing shows the remaining stem of a tree fork. It must be shown now that the force flow in the remaining part is diverted around the wound of the stem of which the fork was deprived. Practical proof has been furnished by observing a 22-mm ⌀ circular notch which was punched into the tree just beside the wound. It took only 1 year for the notch to heal into the state outlined. CAO computer simulations by Volks (1990) have shown that the wound spindle completely coincides with the direction of the force flow. Punching must be done most carefully and in a skillful manner. A hole punched too close to the neighbouring large wound will never heal on the side facing the wound, and may even shrink slightly as it dries up due to insufficient sap. If one punches the hole too far from the wound in order to avoid this shrinking, one moves out of the range of force flow diversion. In this case, the vertical flow of forces will cause the spindle to point in the vertical direction, and will turn it into an ordinary branch hole which has lost the special attraction of a slanted wound spindle. The large wound caused by removing one of the fork members will heal into a vertical spindle corresponding to the main vertical force flow, which leads to highest notch stresses at both sides of the wound. Of course, these notch stresses are not disturbed significantly by the small punched hole beside the wound.

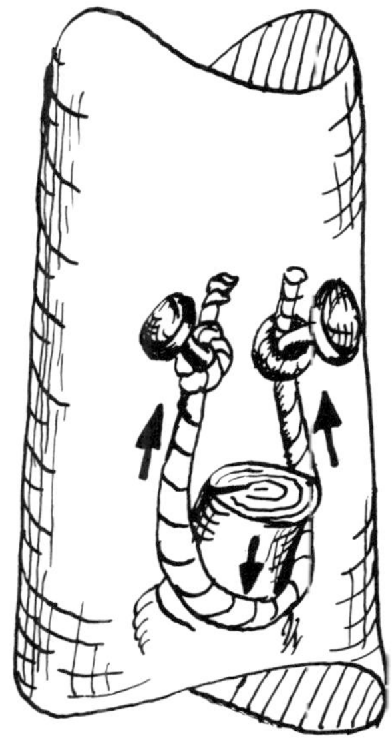

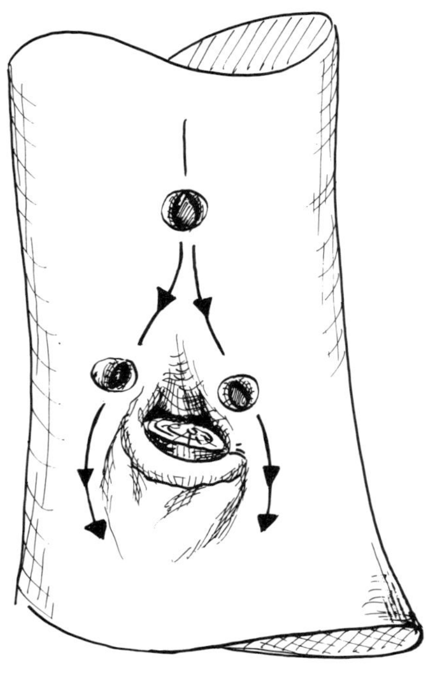

THREE WOUND SPINDLES INDICATING THE DEVIATION OF FORCE FLOW

Three circular notches were punched into a stem around the remaining stub of an old branch. In each of the punched holes, stress-controlled wound healing forms a spindle-type residual opening. All three wound spindles are in alignment with the force flow, and allow one to easily deduce its path around the old stub (arrows). Especially on the observer side, the interestingly marked bulge around this stub can only form if the tree bends towards the plane of the page and the remaining stub is pressed into the loop formed by the bulge. Hole bearing is the technical term used for this load case, which is shown in the simplified sketch above. The rope restrains the stem, which leans away from the old stub.

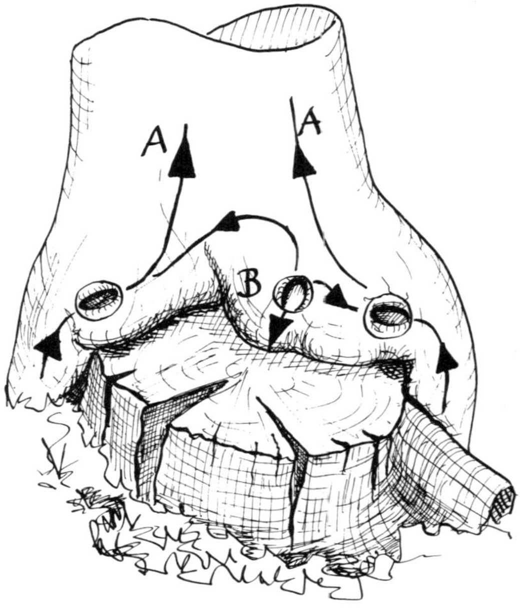

COMPLEX DISTRIBUTION OF FORCE AT THE BUTT OF A CHESTNUT

This chestnut tree is growing around an old stump. Three circular notches were punched into the callus growing around the stump in order to obtain reliable information on the force flow direction. The wind loads indicated by the force flow arrows do not act simultaneously. Load case (A) applies if the stem is bent perpendicularly to the plane of the page, load case (B) if it is bent out of the plane of the page towards the reader. In the latter case, the callus covering one-half of the stump in an elephant-foot manner is pressed against the stump, and stimulates the formation of the vertical central wound spindle. Part of the force flow runs along the length of the callus, and contributes to wound healing of the right- and left-hand side holes before reaching the roots. If the tree is bent away from the stump as in load case (A), callus and roots at both sides of the stump act like tension cords stabilizing the tree. No force flow goes through the callus region (B). It is diverted laterally instead, and stimulates the development of slanted wound spindles. Since these lateral spindles are exposed to both load cases (A) and (B), they bear higher stresses and heal faster than the vertical centre notch which is only exposed to load case (B). The fact that the lateral notches are located in a force flow diversion region, while the centre notch has a mechanically less critical location, may serve as an additional explanation. The orientation of each spindle is in good agreement with the expected direction of the force flow, i.e. with the direction of the principal stress trajectories.

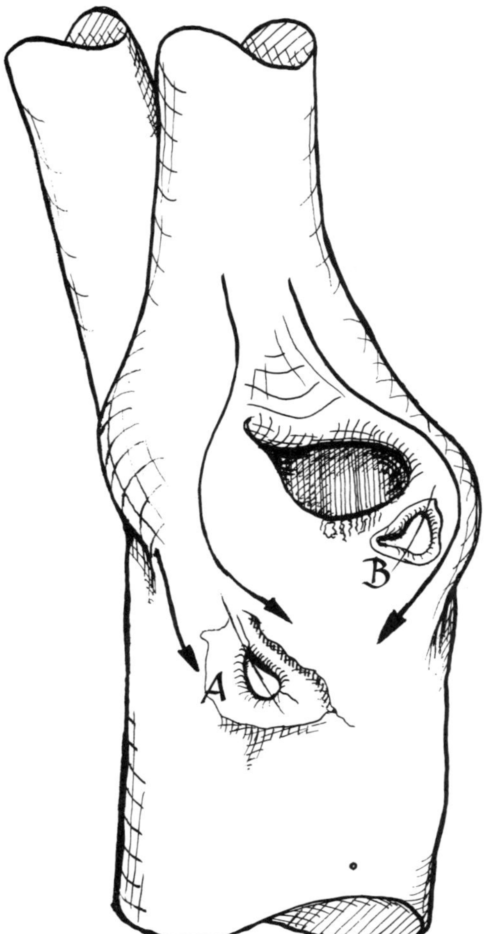

WOUND SPINDLE FORMATION PARALLEL TO THE FORCE FLOW IN A LONDON PLANE
(after a photograph by W. Eisele)

The main stem of this plane was destroyed, and two huge side branches remained one upon the other, inducing high bending stresses at their joints. Thick bulges developed at both sides of the wound left by the main stem to restore the constant stress state disturbed by its loss. Computer simulations by Mattheck and Korseska (1989) revealed that the force flow is distributed as indicated by the arrows. Advantage was taken of the fact that the tree was already injured at two other points (A, B) just in the region where the force flow slants relative to the vertical stem. The drawing shows that the wound spindles of these two partially healed wounds are adapted to the force flow. They heal fastest at the points of highest notch stresses, which are located at both their sides in symmetric alignment to the slanted run of the force flow. It was this interesting chance experiment which first motivated the author to use the "punched-hole method", or rather the "wound-spindle method" to determine systematically the direction of the force flow even under complex loading conditions.

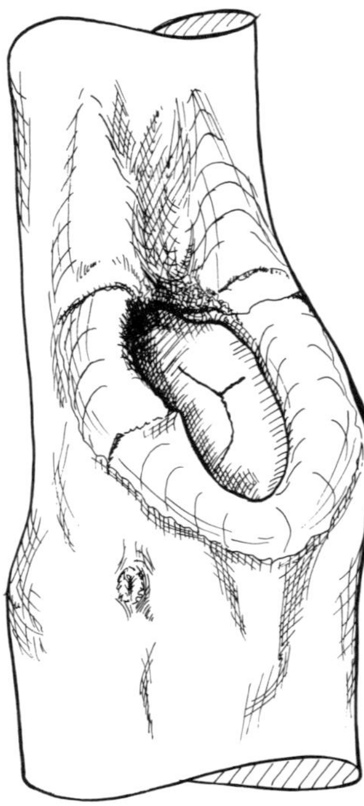

BULGE FORMATION REDUCES NOTCH STRESSES

Wound healing may sometimes be delayed by the persistent stump of an old branch, or an oversized wound which takes more time than usual to close. Trees can activate an additional or even alternative mechanism, which helps to reduce the considerable notch stresses around such slow-healing wounds. According to computer simulations by Mattheck (1990 a), this stress relief mechanism induces the development of axial bulges where notch stresses are highest. The smooth and tightly stretched bark is indicative of the high stresses to which these axial bulges are exposed. Regions carrying minor stresses like that between the bulges above the stump are recognized as such by their dark, crumbly bark. The bulging part is tight and smooth because the old branch inside the tree resists the effort of the remaining stem to bend into the plane of the page. The stub presses from within against the lower part of the bulge. The latter forms a U-shaped sling around the stub of the old branch. This sling stabilizes the main stem, and is under high tensile loading. The space between the ends of the bulging sling directly above the stump is scarcely loaded. The constant stress state has not been reached yet. Bulge formation, however, is indicative of the tree's continuous effort to equalize differences and restore the state of constant stresses, which was disturbed by the loss of one branch. A similar behaviour of bulges forming a sling around the stub was shown in a previous example.

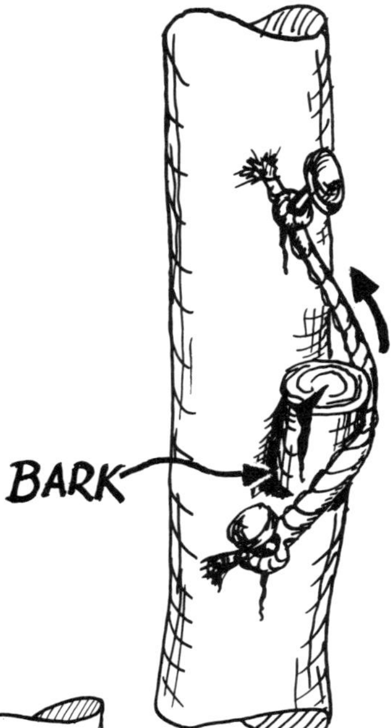

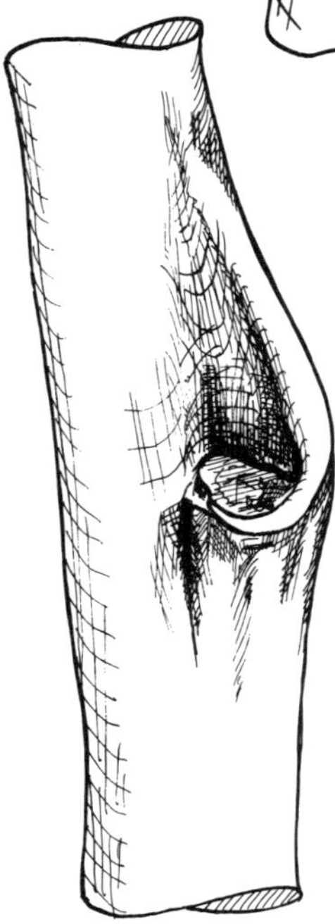

EXTREME BULGE FORMATION AROUND AN OLD BRANCH STUB

The axial bulge shown in the drawing was formed due to unilateral force flow deviation after the right part of the fork had been sawn off. The bulge diverts the force flow in a tension-rope manner around the stub of the missing branch. Since the "rope" also has a remarkable thickness on the right-hand side of the stub, i.e. far from the remaining partial stem, it looks like the old stub is resting in a loop or a sling. Unilateral bulge formation suggests a certain tilt of the remaining partial stem out of the plane of the former tree fork. Another explanation of this non-symmetric bulge formation could be that the fork, which was obviously a compression fork (defined earlier) has unified, i.e. fused tree rings only at the right-hand side of the contact area. Only these fused tree rings are now able to bear tension. The other, unwelded side of the contact area still has bark between the contacting stem parts. No tension can be transmitted there. This could be another reason why the rope-type bulge was formed only at the right, tension--bearing side.

AERIAL ROOTS GROWING AROUND A SCULPTURE

Mechanical contact is known to cause contact stresses (Hertzian stresses) on the surfaces of trees. Contact stresses are compensated by enlargement of the contacting area, i.e. loads are distributed over a larger area to reduce them (cf. a cushion, which distributes contact stresses as one sits on it). Trees grow around the contacting foreign bodies, integrate them, and start to shape-optimize the whole compound structure. In the present case of a head sculptured out of stone and surrounded by the living tree, an additional effect was reached. The living tree seems to animate the dead stone, and tends to accept it as a member of the tree structure. This unique, unusual tree-and-stone complex suggests the idea of a spirit invited to live in the tree to keep away unwelcome threats and visitors.

REMNANTS OF A CRUCIFIX INTEGRATED IN THE WELDED FORK OF A TREE

This birch holding a crucifix between one of its forks was found in southern Germany. The combined effect of contact stresses and natural tree growth led to a rapid integration of the sculpture (note the bulging overgrowth covering the right side of the head). The author was told that integration had been progressing so fast that chiseling was the only way to free the body from the wood surrounding it. This obviously perfect form fit gives proof that adaptive growth can master even complex foreign contours, and integrate the most irregular bodies until all contact stresses are equalized. Regarding all these efforts made by the tree, one should consider its limited alternatives compared with ours: A person touched accidentally by another person in a train or in the subway may easily step aside to avoid contact and contact stress; a tree can neither walk nor move when touched by a foreign body! If trees did not integrate the bodies contacting them, long-time wear could not be avoided. Growing around and integrating foreign bodies seems to be the best alternative for trees to cope with such situations.

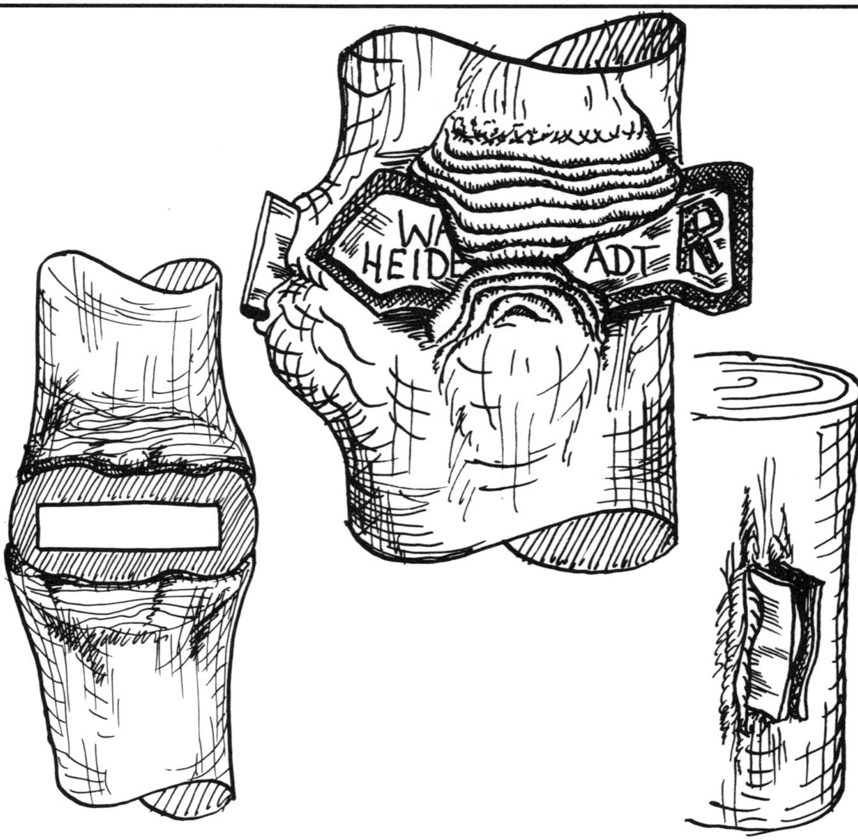

INTEGRATION OF ROAD SIGNS
(Professor H. Ruder, Tübingen, and Mr. F. Dietrich, Karlsruhe, kindly showed me these phenomena)

These road signs attached to a tree induced the formation of tree rings in the narrow gaps between tree and sign. As usual, the resultant contact stresses were reduced by contact surface enlargement, i.e. a local increase in tree diameter. The strange, wavy callus contours must have formed when tree rings stretching for support on the foreign contact surface found only a thin rim for them to rest on. Lateral broadening, level to the road sign, is the natural result of contact surface enlargement of the kind found with foreign stones or in compression forks. The road sign on the left will be completely integrated as soon as the upper and lower callus edges contact and weld. The centre drawing shows another example of road sign integration. Here, the narrow sign buckled because it was clamped in the middle and pushed in the direction of the viewer (out of the plane of the page) by intense formation of wood. The drawing on the right shows nearly complete integration with only the ends of the sign remaining visible.

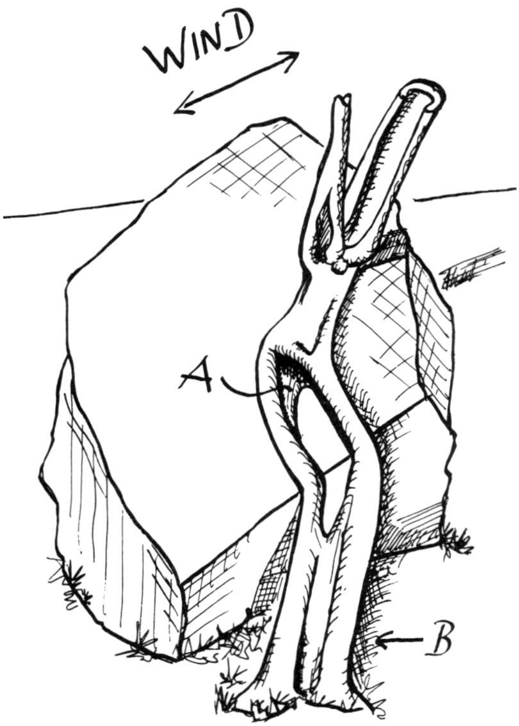

A FLAT STEM RELYING ON A ROCK
(after a photograph by Marita Mattheck)

The enormous flattening of this Turkish olive tree near the edge of the rock can be called an ingenious adaptation in the sense of the mechanical partnership of tree and stone. The contact surfaces are enlarged to reduce contact stresses. This regulative enlargement is particularly marked in the upper part where stems fuse into one. The filled gap in between (A) suggests that this area bears highest contact stresses. Wind blowing along the edge of the rock may lead to bending (arrows). This load case is mastered by an I-bar or 8-shaped profile (B) formed at the very butt, where bending loads are highest. Such profiles were introduced as being ideally optimized, uniaxial bending structures (roots of spruce), whose properties are also appreciated in civil engineering. Possible bending of the stem towards the stone is compensated by the support of the stone itself at (A). Bending away from the stone should be negligible because the tree is at the lee side of the rock. However, this bending would be absorbed by the gravity of the tree, and would only effect a short-term decrease in pressure at (A). It is evident that the tree-stone complex is fully adapted to all possible load cases. The tree completely relies on the stone, and profits by its partnership. Should the stone be removed at a stage of perfectly optimized support, the tree would be doomed to deteriorate and become a faulty design facing imminent breakage.

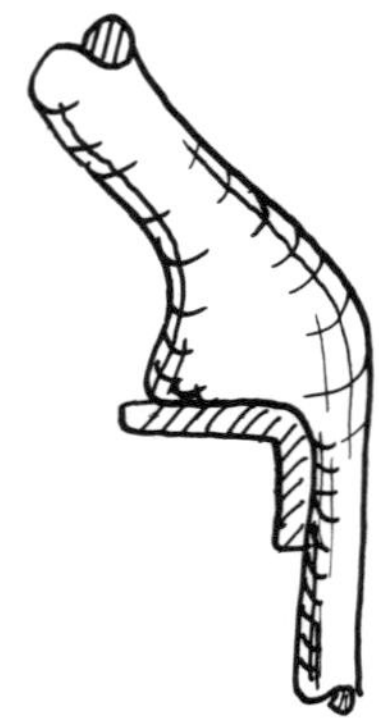

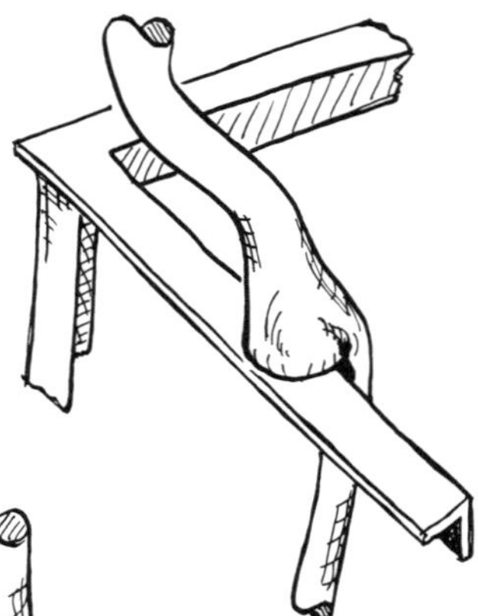

AN ELEPHANT FOOT SET ON A RAILING

In the botanical gardens of Singapore, the heavy branch of a mighty bush was seen leaning over the sharp edge of a steel railing. Since no significant winds blew along the edge of the railing, the cross-section of the lower branch remained circular, and did not ovalize or even form an 8-shaped profile like the flat stem dealt with previously. The main load seems to be the gravity pressing the branch against the edge. A drastic reduction of these highly localized contact stresses is achieved by the formation of a pad, which strikingly resembles an elephant foot (in fact, nature has shaped the feet of elephants so as to meet a similar requirement, i.e. to bear the high loads of axial compression). Furthermore, it is interesting that the diameter of the stem is much smaller below the contact zone than above it. Obviously, the support of the railing considerably relieves the stresses in the lower part of the branch. In this case, too, the branch would break if its mechanical partner was removed. Adaptive growth would not be fast enough to compensate such abrupt mechanical changes.

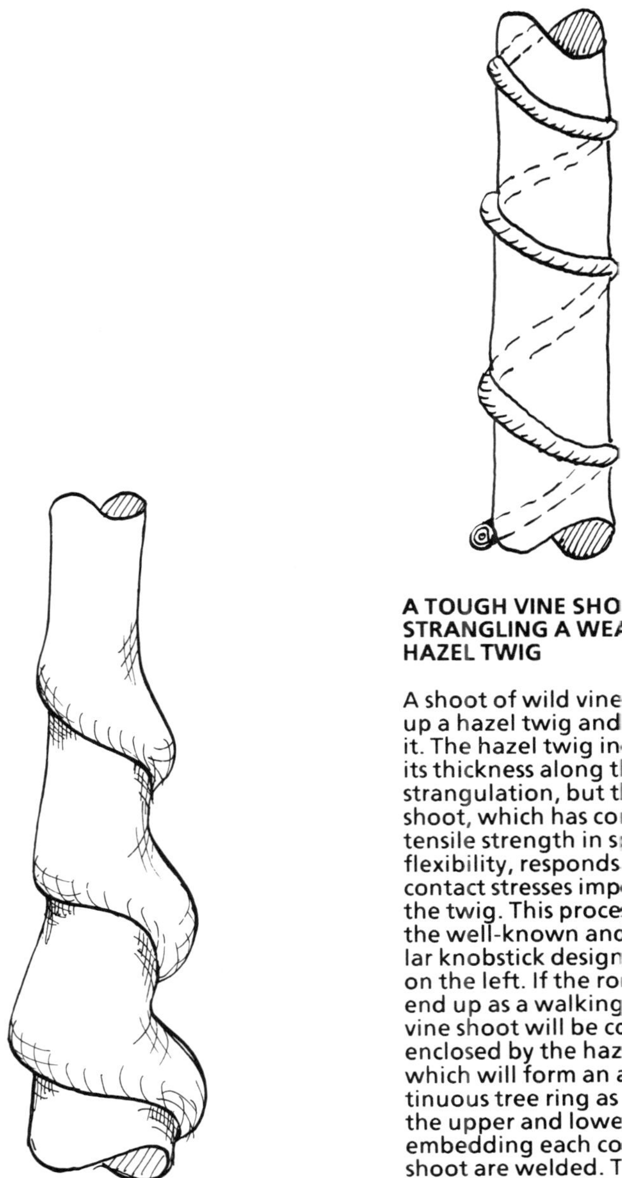

A TOUGH VINE SHOOT STRANGLING A WEAK HAZEL TWIG

A shoot of wild vine climbs up a hazel twig and entwines it. The hazel twig increases its thickness along the line of strangulation, but the vine shoot, which has considerable tensile strength in spite of its flexibility, responds by high contact stresses imposed upon the twig. This process results in the well-known and very popular knobstick design illustrated on the left. If the rod does not end up as a walking stick, the vine shoot will be completely enclosed by the hazel twig, which will form an axially continuous tree ring as soon as the upper and lower bulges embedding each coil of the shoot are welded. This tree ring enables the hazel bush to survive, and prevents it from strangulation.

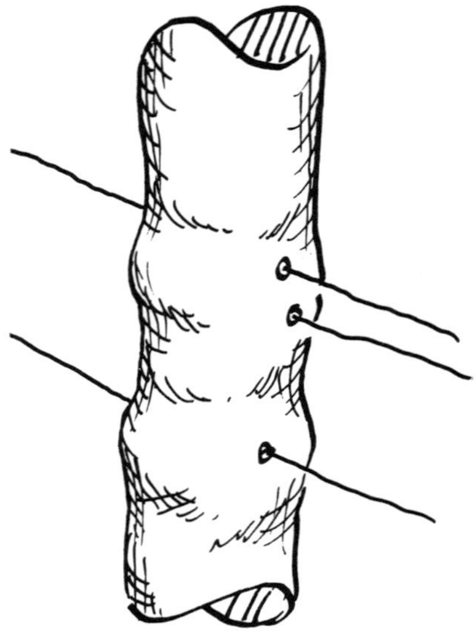

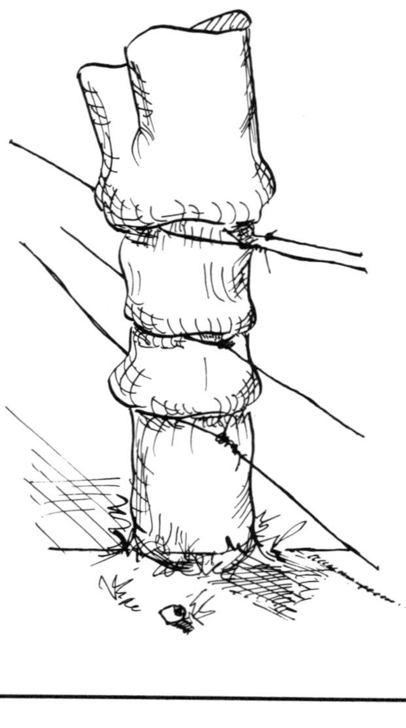

STRANGULATION BY FENCE WIRES

This tree was seen serving as a pale for wires fencing a pasture in southwest Africa. As the tree increases its thickness, high contact stresses develop on the surfaces in between. Like the vine shoot in the previous example, the wire becomes wrapped more and more by the tree. New continuous tree rings are formed, which bridge the gaps between the rings of wire, and cover them up by axial welding. The remaining bulges will smooth in the course of time. Passersby will see a wire running out of the tree, and will wonder whether it forms part of a loop inside the stem, or whether a bore was drilled for it to run straight through.

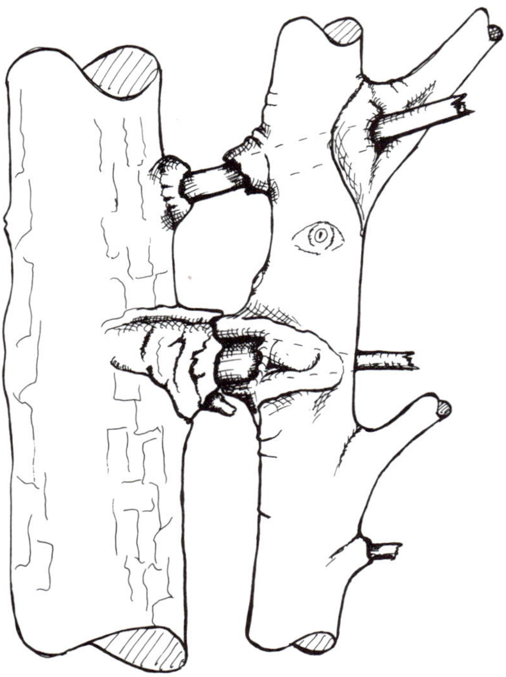

TREES OF DIFFERENT SPECIES ARE TREATED AS STONES
(Mr. D. Polle, Karlsruhe, kindly showed me this tree)

Two branches of the pine tree on the left came into contact with the red beech on the right. Wind caused the pine branches to wear the bark of the beech. Wound healing was by partial form fit followed by enclosure of the barkless contacting branches. The notch stresses around the edges of the new joint are limited by collars which could deceive one into believing that the beech accepts the pine and fuses with it. However, since different species neither develop common cambium nor common tree rings, it is only a simple form fit, suggesting that the pine is treated as a lifeless foreign body. The trees do not even weld in the middle of the lower joint where they are rather intimately connected through bulges coming from both sides. Their relationship resembles a separation-of--property marriage. They live together with comparable burdens and loads, but strictly separate nutrient and water supplies. For the time being, they both take mechanical advantage of their common framework structure, which, however, will be destroyed as soon as the pine branches decay and leave large holes in the beech, which is already exposed to decay in the contacting areas at the present stage.

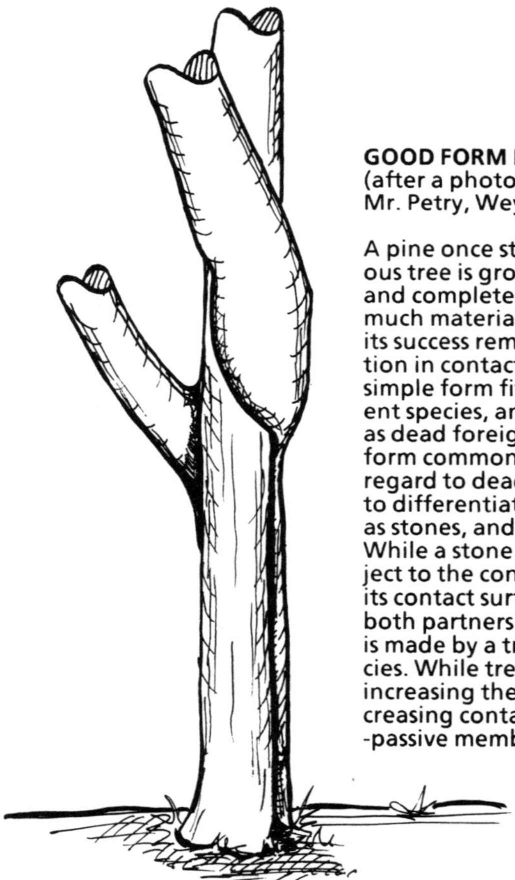

GOOD FORM FIT WITHOUT WELDING
(after a photograph by Mrs. and Mr. Petry, Weyher, Germany)

A pine once standing behind a deciduous tree is growing around the latter and completely entwines it. Although much material was invested in this union, its success remained limited to a reduction in contact stresses and a subsequent simple form fit. The trees are of different species, and treat one another as dead foreign bodies, which do not form common tree rings. However, with regard to dead foreign bodies, one has to differentiate between objects such as stones, and trees of a different species. While a stone keeps its shape and is subject to the contacting tree which enlarges its contact surface by adaptive growth, both partners are active when contact is made by a tree of a different species. While trees assist one another in increasing the contact surface and decreasing contact stresses, stones are ever-passive members.

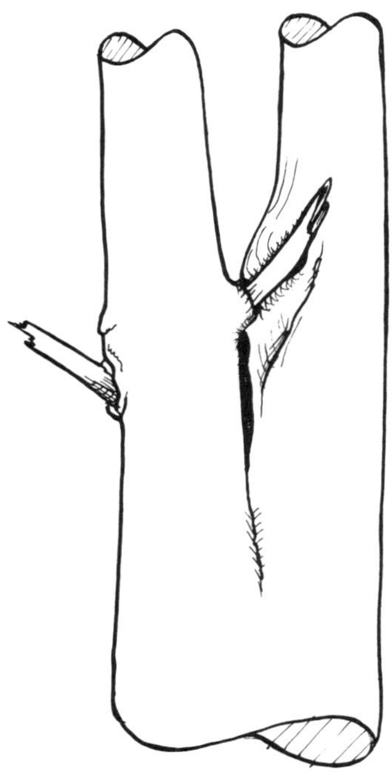

DEAD BRANCHES FROM THE SAME TREE ARE ENCLOSED LIKE STONES

Coming from the left stem of a narrow tree fork, a protruding branch contacts its twin on the right. Wind--induced wear and decay have removed the bark of the branch, and turned it into a dead member without any cambium left for fusion with the stem contacted. The right stem is still stimulated by the mechanical contact of the dead branch, and proceeds to envelop it. The stump is surrounded by a typical callus collar at its base. It will soon break at the notch-edge of this collar, and leave a branch hole as well as the two bulges into which it is embedded. Only adaptive growth will heal and smooth the two marks in the course of time until they seem to be one with the stem. The example shows that the response of adaptive growth to localized stress peaks is also a localized one. It is a non-anticipative response due to lack of signals or indications which could induce the stem to stop embedding the branch awaiting its decay. The tree proceeds as long as it is stimulated. It adapts to all kinds of mechanical impacts but cannot adapt to future requirements and situations. If the contacting branch was still alive, the stem would probably weld with it. In this case, a mechanical frame would be formed which would be of greater mechanical advantage to the tree in saving material.

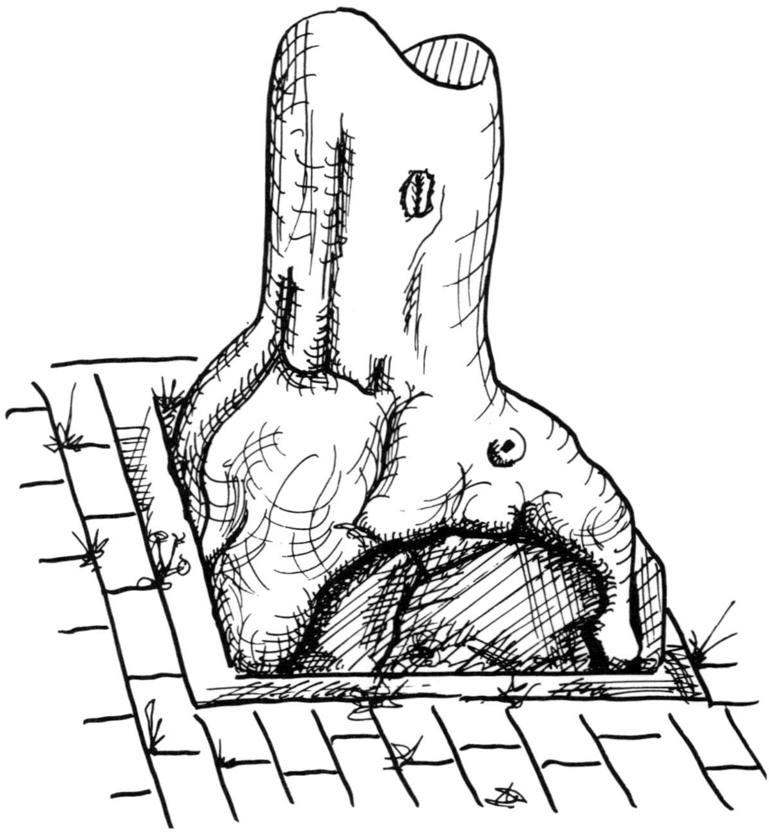

A STONE TRANSMITTING PRESSURE INTO THE SOIL

A city tree bricked up in a cobbled pavement is growing over a large stone. Although this structure looks unstable and uncomfortable, it depends on the wind direction whether it really is or not. If the wind blows out of the plane of the page, the tree base facing the reader is pressed against the stone, which transmits this pressure into the soil just like the root stock would. If the wind is blowing perpendicularly into the plane of the page and bends the tree away from the reader, the tree base is lifted off the stone. Since the contact surface can only transmit pressure but not tension, the stone cannot stabilize the tree in the latter case. The tree shown in the figure is facing a nearby house. Since the stone is on the lee side, i.e. with no tension loads occurring, it can survive with this pressure-transmitting partner.

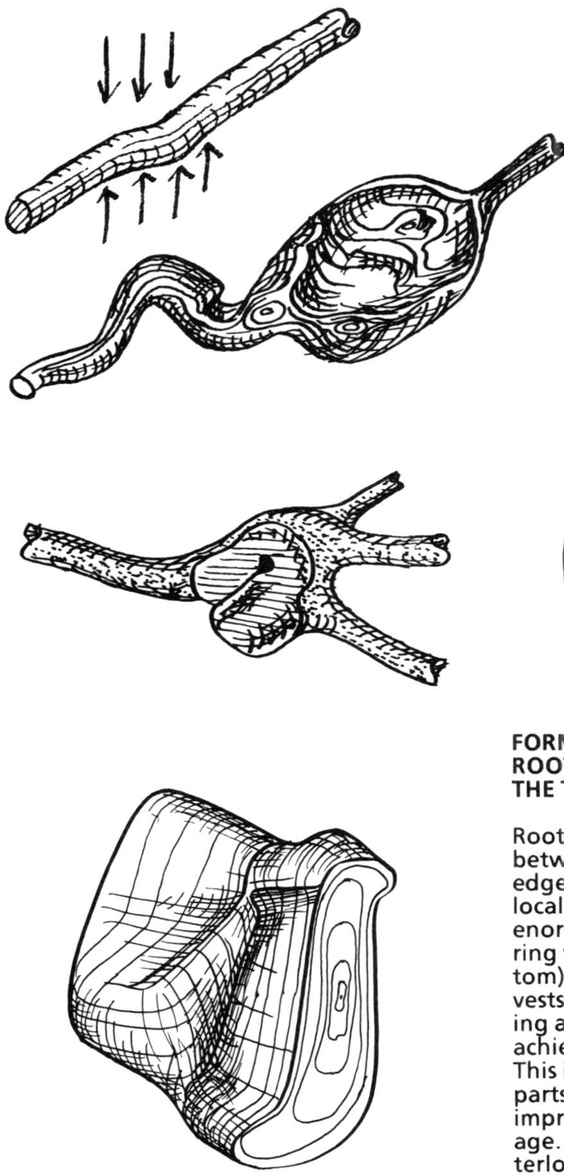

FORM FIT OF EMBEDDED ROOTS HELPS TO ANCHOR THE TREE

Roots may often get trapped between two stones whose edges and corners cause high, localized contact stresses. The enormous differences in tree ring widths (figure at the bottom) show that the root invests much material in growing around the stones and achieving a good form fit. This intimate contact with parts of the soil considerably improves the tree's anchorage. The root-and-stone interlock has the function of a barb or knot retaining the tree.

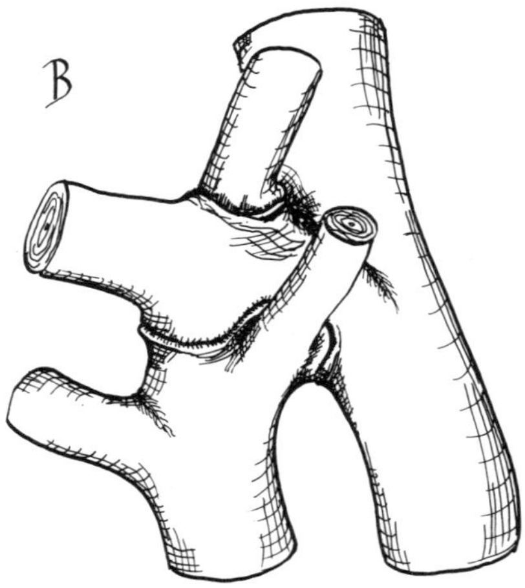

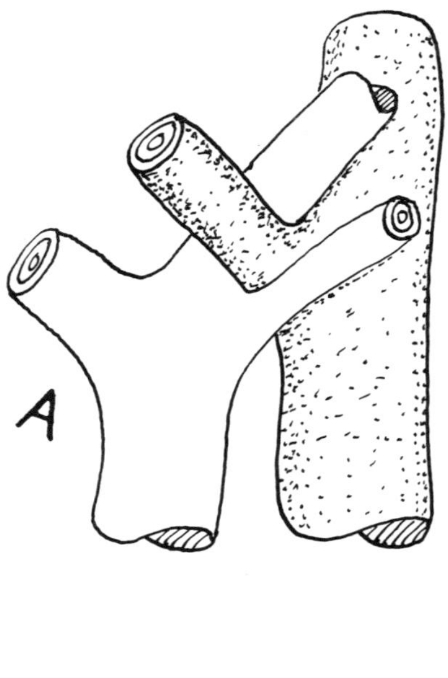

TWO WELDED HORNBEAM FORKS

The sketch below (A) shows the assumed initial state of two hornbeam forks hooked together. It seems to be ideal for a rapid fusion of these two components. While each fork is shape-optimized according to its own needs when single, loads change as fusion starts to cause high contact stresses which are further increased by wind loading. The contact surfaces are enlarged by adaptive growth, i.e. the forks grow around each other, and all sharp edges and corners (notches) are smoothed in order to avoid any kind of stress concentration. Figure (B) shows the present state of a tight form fit reached by suppressing all relative movement of the forks. Although hidden from the eye, one can be sure that the tree ring arrangement is optimized, too, to avoid shear in between, and delamination. The forks are already fused at some points, and at other points are just before welding. Therefore, the outermost tree rings are probably arranged perpendicularly to the interface as is neccessary for successful fusion.

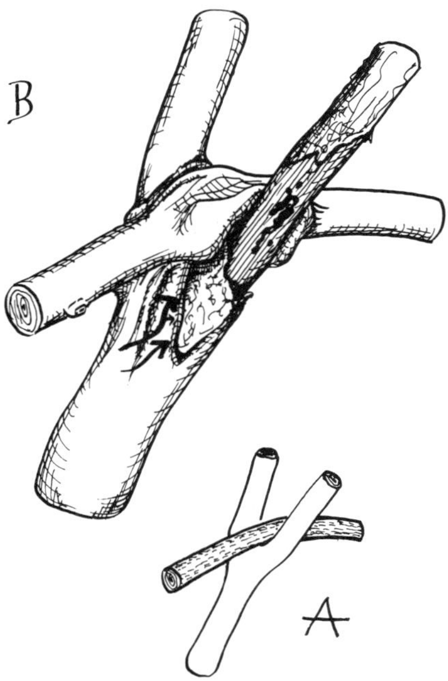

PARTIAL WELDING - PARTIAL FORM FIT

The branch and fork shown in (A) are both of the hornbeam species, which, like also red beech, is known for its disposition to fuse in a sophisticatedly shape-optimized way. The right part of the fork has lost its bark. Its decay and death might have been caused by the very contacting branch injuring it in the initial state (A), or by impacts which affected it before contacting. The left part of the fork is still alive. Although both parts seem to be in the same position and are, in addition, both surrounded by the bulging contacting branch adapting to them (B), they are treated differently. The contacting branch completely fuses with the left part of the fork, and a common tree ring is formed, which welds them. The dead and wormy right part is treated as a dead body with no other status than a stone, for example. It is a branch without cambium and, therefore, cannot be accepted as a partner for common tree ring formation and welding. The collar (arrows), which is in the process of forming at its very base, indicates that the tree is preparing itself to lose the decaying part, and to close the remaining hole by wound healing. The right-hand contact surface enlargement will appear to have been a wasted effort, an error the branch made due to the lack of a signal announcing the death of the right branch. Once again, adaptive growth proves to be a non-anticipative process which only responds to the dictates of current mechanical stimulation.

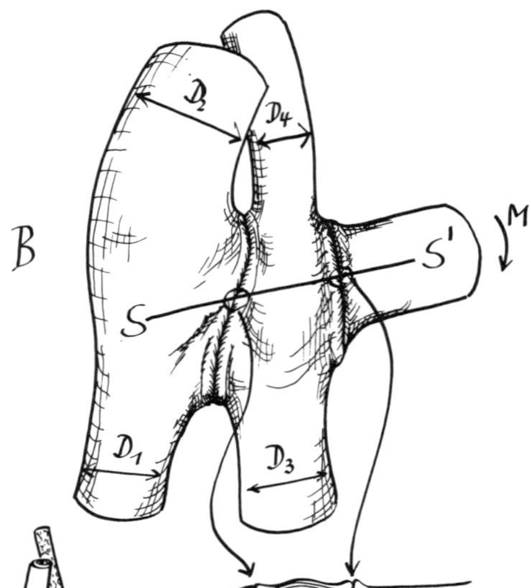

INTERESTING CHANGES IN DIAMETER

Figure (A) below illustrates the situation before the fusion of two hornbeam branches. In (B), the smoothed notches in the contact zone and the internal arrangement of tree rings along the force flow (S - S') indicate that the process of welding is nearly completed. Welding has curiously changed the diametrical relations throughout the compound structure. The vertical, left part of the fork is narrower now below the bifurcation ($D_1 < D_2$) than above it, and the adopted branch on the right has just the opposite relations ($D_3 > D_4$). The latter must have had a nearly uniform thickness ($D_3 \approx D_4$) before contacting the fork. However, it is exposed to gravity and wind bending moments (M) after welding, and increases in thickness below the contact zone ($D_3 > D_4$) on account of considerable, additional loads. The left, vertical part of the fork must have been thicker below than above ($D_1 > D_2$) before welding. The force flow, which has made the contacting branch increase its thickness in the lower part, is obviously missing in the left stem (D_1) because of the frame effect achieved by the left part of the fork and the adopted branch supporting each other below the welding. This might explain both the diametrical relations $D_1 < D_2$ and $D_3 > D_4$, respectively. The left, upper part has become the thickest part of the structure because it cannot participate in the supporting frame effect. It probably has more leaves and a larger crown than the right, upper end marked as D_4. This might explain the difference between them. A walk through a young hornbeam forest is a worthwhile pastime which may give deep insight into the broad spectrum of adaptive growth phenomena governed by the constant stress hypothesis.

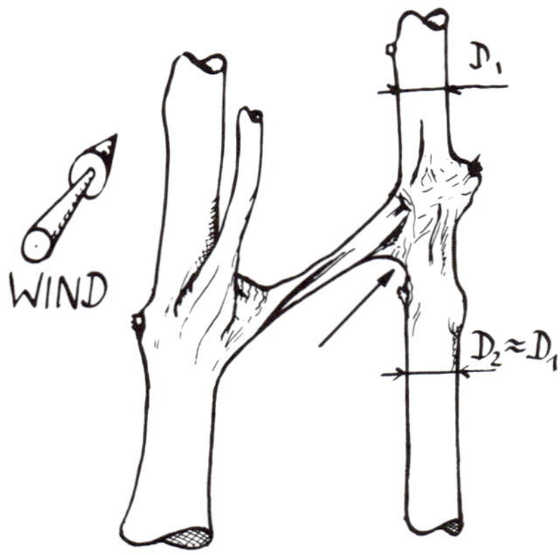

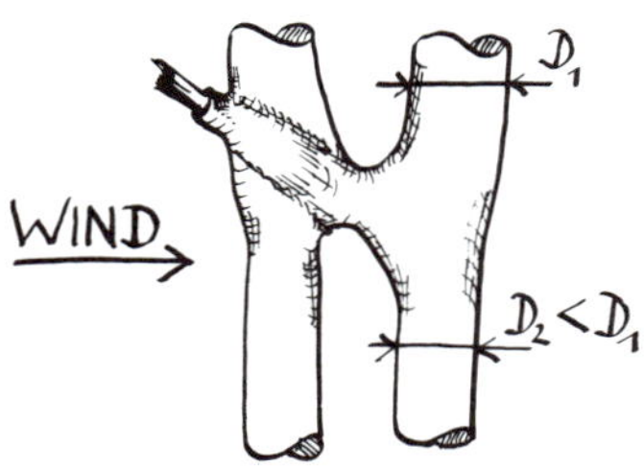

H-TREES: BEAMS BECOME FRAMES

Each of the tree pairs above and below was formed by one branch of a stem extending to touch the neighbouring tree, which is of the same species and possibly even shares its root stock with the adjacent one. Again, mechanical contact is followed by a stress-reducing enlargement of contact surfaces and integration of the contacting branch. Common tree rings are formed as soon as their internal arrangement allows this. Final welding results in similarly H-shaped, yet altogether different pairs of trees. The maple trees at the top are characterized by an almost equal thickness of the right-hand stem above and below the new joint connecting the adopted branch. In the case of the chestnut pair shown below, both stems are much thicker above the bridging branch than below where the tree components support each other and are reinforced like in a frame. These strange diametrical differences, in spite of similar fates, can be explained as follows: Due to their position within the stand, the upper trees were loaded by wind almost exclusively perpendicularly to the plane of the frame (plane of the page). Loaded in such a way, the frame has hardly any reinforcing effect because the individual stems of the framework do not support each other. Both stems swing out of tune and out of the H-plane. The connecting branch is loaded by torsion and bending. It twists like a rope, and, in addition, smoothes the notches at the connecting joint (arrow). The diameter-to-height relations of the maple pair can be called normal. The pair of chestnut trees shown below is mainly loaded in-plane, which means that the stems can take full advantage of the frame design. The height--diameter relations deviate from normal because the stems support and relieve one another below their smooth-jointed interconnecting branch ($D_1 > D_2$, slower thickness growth), but are on their own above the bridge and grow thicker. However, both types of H-trees have perfectly smoothed notches at their connecting joints. They are stable without wasting too much material. The H-tree example illustrates the role and range of the constant stress hypothesis.

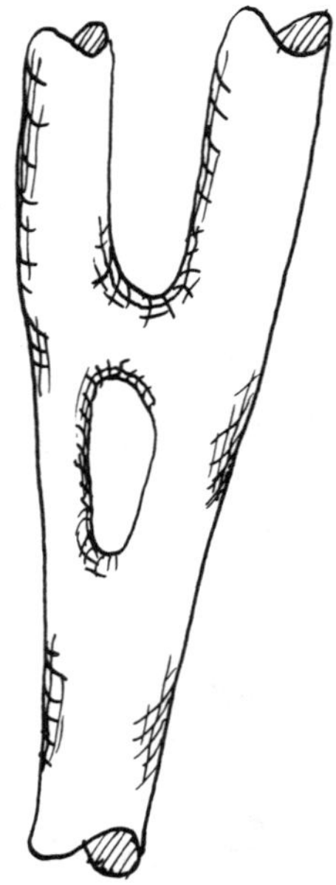

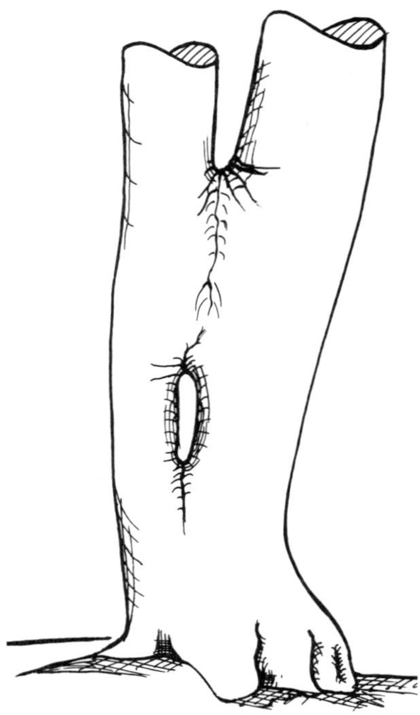

COMBINED MASSIVE WELDING AND DIAMETRICAL ADAPTATION

The left and right partial stems of a common root stock and of a near--to-ground tree fork, respectively, are welded over a length of approximately 1 m, each. Below these fused sections the members are supported by the structures' reinforcing frame effects. Their increase in thickness is negligible compared with that of the stems above the welding. The upper and lower notches, i.e. the holes below and the notches of the new forks above can be assumed to be shape-optimized to avoid localized stress peaks and to perfect these balanced designs. The diameter relations indicate that the main direction of wind loading is in--plane of the frames grown.

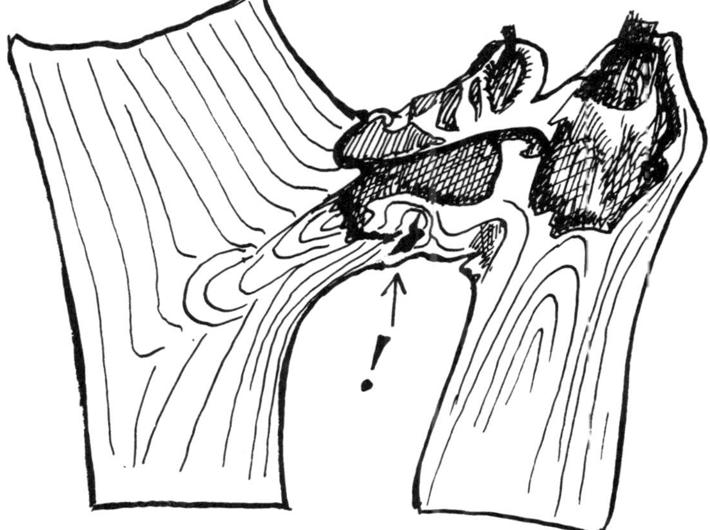

DEATH OF ONE PARTNER AFTER WELDING

Two partial stems sharing one root stock contact, and are pushed by geotropic reaction wood formed on the part of the (missing) outermost stem to join in a perfect form fit. First common tree rings (!) show as one saws through it, and shape-optimized, well-rounded notches confirm that it is well-prepared for the future. However, its efforts, material, and energy are wasted and seem to have been in vain the moment the right stem breaks off. Its failure was probably caused by careless transport of logs from a near-by logging road. It was predetermined, so to speak, by its disadvantageous position. All that remains are the traces of an adverse fate which ruined the product of a successful shape optimization. Adaptive growth will be left with the task of healing the wounds of this unkind damage. The nearly closed wound at the top of the right-hand side "elbow" shows the first success of these restorative activities.

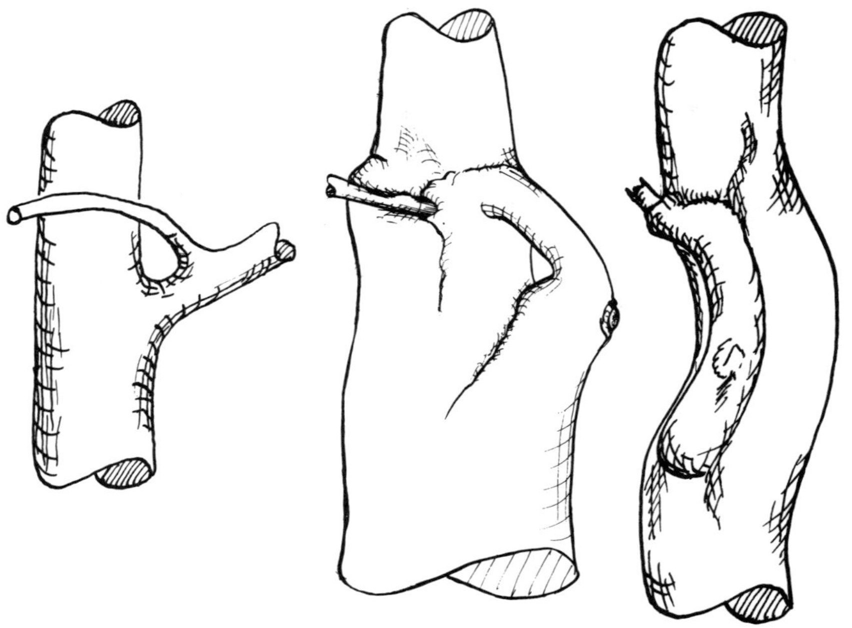

A "JUGHANDLE" BEECH

When young and flexible, one of the branches of this beech was bent towards the supporting stem, perhaps in stormy weather. Most probably, it got caught on some roughness of the stem or on another branch, and was pressed against it by its own elasticity. It became completely enclosed by the stem, and welded into the curious jughandle shape shown in the middle. Thus, sap transport along the handle was interrupted, and the tip of the branch broke off. Its fate is completely different from that of the integrated part. While the dead tip is treated and embedded like a contacting stone, and is bound to decay due to its inability to produce new tree rings and to fuse, the handle itself remains a living member of the tree. The beech has to take the consequences, i.e. share its down-stem force flow (e.g. wind loads entering in the upper part) with the new member. Participation of the handle has a stress-relieving effect on that part of the stem between the two joints. In accordance with the constant stress hypothesis, the tree rings produced in this protected area are thinner in order to restore the state of constant stress along the stem. The side view on the right shows the concave contour of the stem profile at handle level. Although the tree gains little advantage from the handle, it makes the best of this unusual design by systematic shape optimization and minimization of the extra risks of failure imposed by its new member.

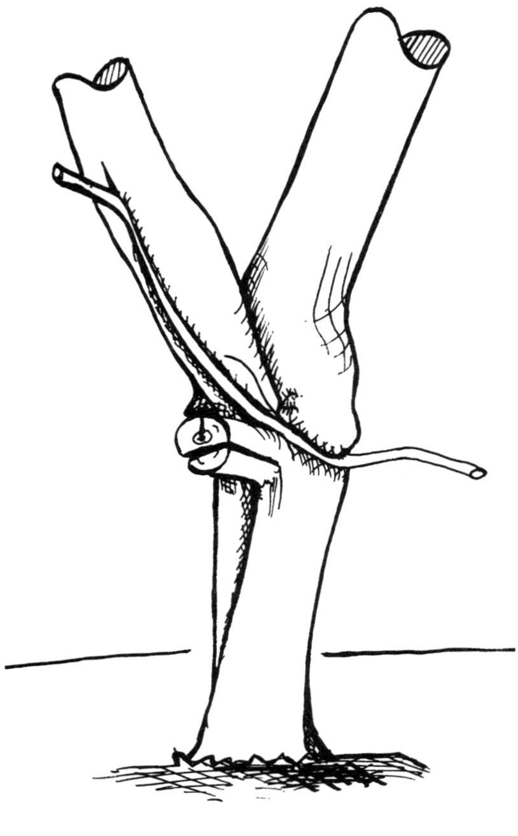

SELF-INFLICTED ENTANGLEMENT OF TWO PERFECTIONISTS

Trees are perfectionists, which constantly adapt and optimize their shape to maintain a constant stress state and survive. Useful as they are in most cases, these self-preserving mechanisms may lead to self-inflicted confusion and growth in excess of the essential. The two neighbouring trees shown on the left are an exemplary case of adaptive chain reactions to the disadvantage of the trees involved. These two trees of the same species are joined, undergoing the usual process of contacting and stress-reducing contact surface enlargement. Growth is moderate until the tree behind comes into contact with a dead stub belonging to the front tree. High localized stresses are caused and lead to excessive growth resulting in new contact areas. As the trees grow around each other level with the dead branch, they are disturbed again by a dead slanted branch alien to the species, which got stuck between the stub and the overgrowing wood. The dead branch is treated as a dead foreign body and is covered up likewise. More and more material is attached in this way and complicates the situation. Cases where adaptive growth causes problems instead of solving them are rare. They are limited to chain reactions of the kind described above, i.e. contact and first adaptive attachment of wood → contact between the wood attached and foreign bodies → new contact areas → simultaneous reduction of the first contact stresses and the new contact stresses as well.

GROWTH STRESSES IN THE AXIAL DIRECTION

Kübler (1959 b, 1983) was the first to develop a straightforward, elegant analytical description of growth stresses in trees, which is still valid and referred to today. Growth stresses are much higher in the axial than in the circumferential direction. They are one of the distinctive characteristics of living trees and decrease rapidly as soon as a tree dies. Growth stresses are internal stresses which can be compared to thermal stresses.They are not related to external loads, but are produced as new wood cells expand in the process of tree ring formation. As a rule, growth stresses, and especially axial tensile stresses on the surface are higher in angiosperms than in gymnosperms. Axial force due to gravity is not regarded here. A thin stem or branch split in two as shown on the right reveals that compressive stresses act near the cut, and bend the segments apart, and that tensile stresses, acting at both their sides, complete the bending effect. Splitting, so to speak, creates two bending bars which separate as a result of internal bending stresses acting in each of them. Trees which are split by storm or lightning behave in a similar way (drawing below). It is easier to grasp the nature and effects of axial growth stresses if one imagines them to be thermal stresses which developed as a result of centre heating and surface cooling processes. The example to follow shows that axial cracking and segment separation due to internal stresses may cause serious felling problems. Because of the axial equilibrium of forces, a compensatory compressive force (left drawing at the top) is acting near the centre of the stem. This leads to a vanishing axial force, resulting as compressive centre and tensile lateral forces are integrated over the cross-section.

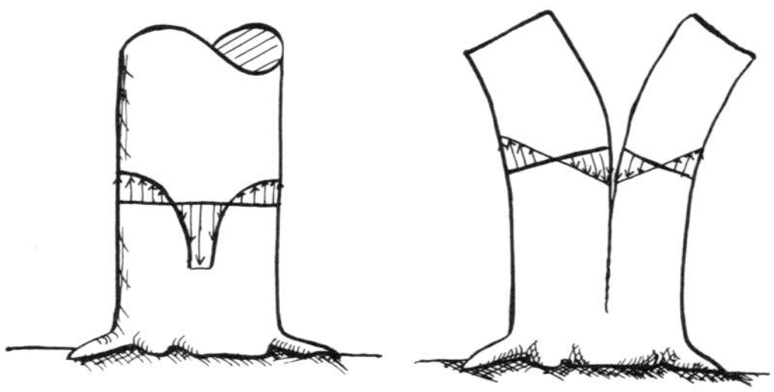

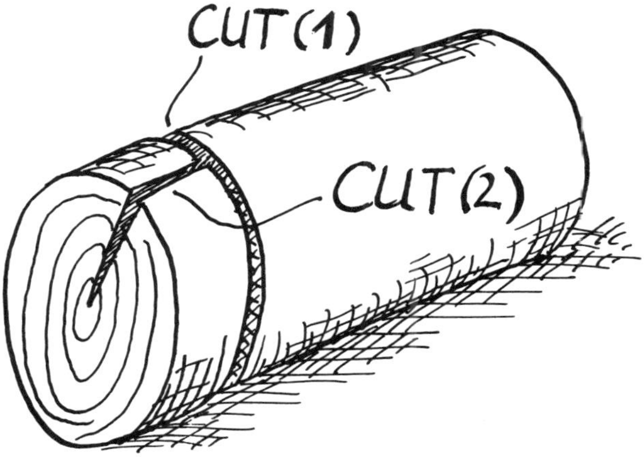

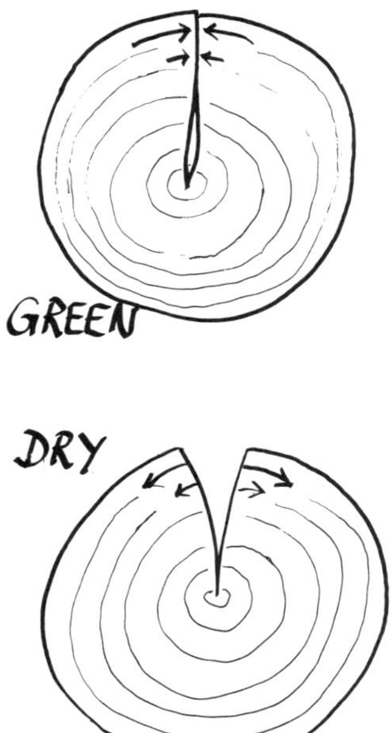

DEMONSTRATION OF CIRCUMFERENTIAL STRESSES

Making a radial cut into a slice freshly sawn from a green tree, one observes as one pulls out the jamming saw that the outer part of the crack closes immediately while the inner one remains gaping. The closing, outside crack is indicative of compressive stresses in a circumferential direction in the tree's surface regions. If one allows the wood slice to dry for a couple of days, the green-tree growth stresses decrease, moisture gradients effect high tensile stresses on the circumference of the slice, and the entire saw cut finally opens wide. This experiment shows that growth stresses, in fact, only exist in the green tree.

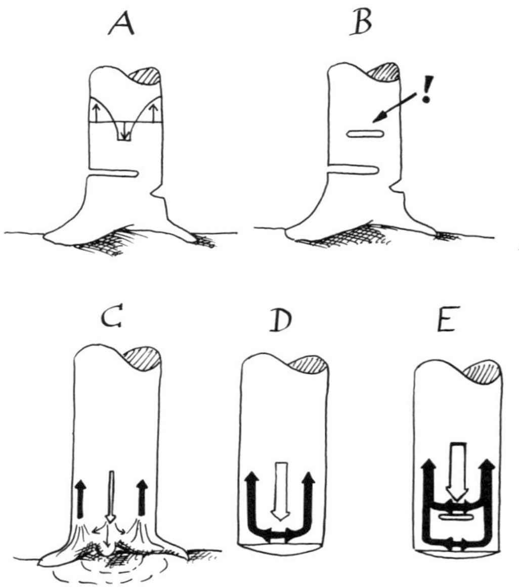

END-SPLITTING AFTER FELLING

When cutting the fine, mature stem of a noble, 160--year-old beech one cannot be but disappointed if large axial cracks are found running along its length after felling. Quick-motion studies have shown that these cracks are not only due to felling impacts as such, but rather may start right after cross-cutting and develop as the tree turns over. It is evident that end-splits such as these are caused by growth stresses in response to disturbances of the stem's axial force equilibrium after cuts of the type illustrated in (A). This cutting technique causes the cross-cut end of the stem to bulge into a convex shape (D). It is the combined effect of axial extension due to compressive stresses in the stem centre, and lateral tension (black arrows in D) due to axial tensile stresses acting on the stem surface instead of entering the roots (C). End-splitting is caused as tensile stresses are diverted along the cross-cut end and turn into lateral tension, which cleaves the very marrow of the stem in the attempt at re-establishing the equilibrium. At the Karlsruhe Nuclear Research Centre a new felling technique (B) was proposed and computer-simulated, the success of which is mainly due to a stress-relieving keyhole cut (!) about one stem radius above the traditional cut. This two-level cut (B) reduces the fatal lateral tension by simply distributing it as shown in (E). The method was successfully field-tested in cooperation with Prof. Dr. Keilen, Landau Forestry Department, Germany. Its excellent results, however, have been due to high-precision cuts made under the supervision of an experienced technical adviser. The complex procedure of determining the exact location and size of each individual stress-relieving notch for each individual tree and each position in a stand makes it a method which is not recommended for general application. A detailed description of the method is given by Mattheck et al. (1989 b).

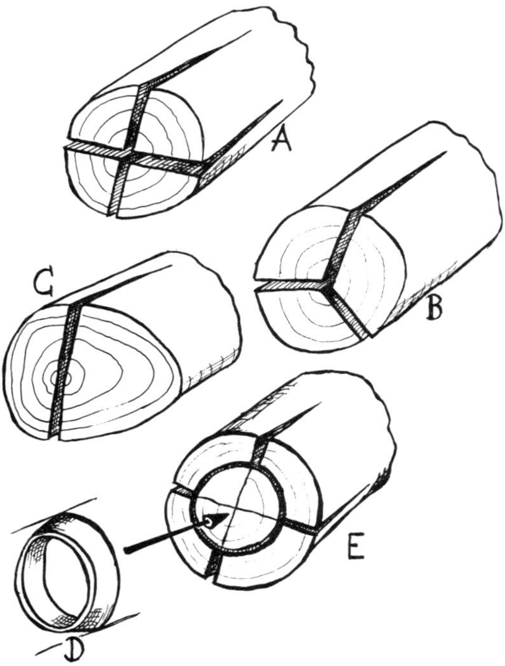

A CONICAL COMPRESSION RING WHICH LIMITS END-SPLITTING

Stems can have different types of cross-cut ends. Circular, even stems show a symmetric distribution of growth stresses, which leads to X- (A) or Y-shaped (B) end-splitting with no predefined direction of crack initiation. Slanted trees, trees on slopes, or trees under unidirectional wind loads tend to ovalize due to unilateral reaction wood formation. Their crack lines are such that they normally split along the smaller semiaxis of the oval cross-section (C). Running through the centre of the marrow, i.e. through the centre of axial compressive stresses, they are perpendicular to the larger semiaxis, i.e. to the long lever arm which finally opens the crack due to the high tensile stresses acting on it. It is reported that eucalyptus trees occasionally crack along their full length, and, split in two, are side-catapulted some distance so that any logger who is nearby is killed by them. The author saw even German beech trees cracking along their full length just after cross-cutting and before hitting the ground. For such highly prestressed trees the "slanted-cut" method described later would be an appropriate causal method to reduce growth stresses *before* cracking. Beeches such as these are exceptions to the rule, and emphasis was placed on the development of a device to reduce cracks in trees with moderate, regular crack formation characteristics. The conical steel compression ring (D) described in detail by Mattheck et al. (1989 c) is driven in with its larger diameter facing the cross-cut end of the log. Since the wider end of the ring seizes more wood than the narrowing end can hold, the enclosed wood is tightly pressed together and possible cracks are closed inside the ring. With no cracks inside, the compression ring fulfills a prophylactic function and will safely avoid crack formation during the drying phase at least inside the ring (marrow). Existing cracks are closed and will be less susceptible to further crack growth during seasoning. This ring, too, was successfully field-tested in cooperation with the Landau Forestry Department, and has been meeting with great interest throughout Germany. However, in spite of all enthusiasm it is admitted that the ring is a "symptomatic therapy" and does not eliminate the cause of cracking like the "slanted-cut method" described next.

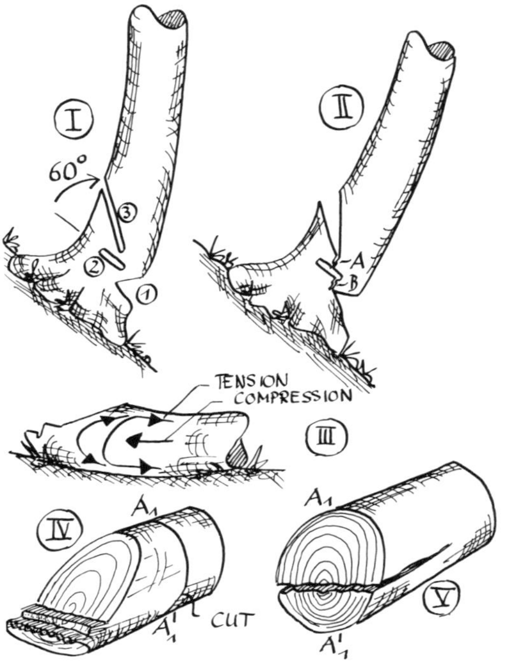

MINIMIZED END-SPLITTING DUE TO SLANTED CUTS

Of all the techniques investigated, the combination (I) of keyhole cut (2) and slanted cut (3) has been yielding the best results. Slanted cuts have more or less the same effect as the stress-relieving keyhole cut described previously. Their goal is to minimize end-splitting due to tensile stresses which act on the surface of the upright tree and are diverted along the cross-cut end during felling. Slanted cuts ensure that diversion takes place at a safer distance from the cross-cut end (III). Trees which are treated in this way hardly show any lateral tension on their cross-cut ends. Computer simulations by Mattheck and Walther (1991) have verified an effective suppression of end-splitting due to growth stresses. In order to largely avoid additional axial cracking caused by hinge bending moments, each slanted cut must be accompanied by a keyhole cut (2) near the felling notch (1). As the tree starts falling (II), a crack "flashes over" from the tip of the slanted cut (A) into the keyhole cut. The hinge immediately narrows, and a second crack (B) as a rule follows suit by flashing over into the felling notch. The tree remains uninjured as it drops to the ground. The combination of keyhole and slanted cut protects the stem from upward cracks originating from the hinge. In the framework of red-beech felling experiments in the Eußer Valley, Palatinate Forest, many of the beeches treated in this way by A. Weiter (Speyer, Germany) were found to crack (V) as soon as the bevelled end was cross-cut (IV) after felling. This, indeed, is convincing proof of the success and value of this combined technique. Provided, of course, that the bevelled end remains, it can be considered the most feasible method recommended for forestry practice.

SPIRAL-GRAIN NOTCHES PROMOTE END-SPLITTING

Spiral grain, which has been shown to be advantageous to the living tree in previous examples, can be a serious drawback if it occurs in straight, high-quality stems marked for felling. Although the twisted ribs and the rope properties help to prevent the tree from failure, they may be its doom once felling disturbs the mechanical equilibrium cherished over a lifetime. In the present case, the notches between the twisted ribs (arrow) seem to attract the cracks originating in the tree centre. Crack growth is accelerated, the crack runs along, and reaches the edge of the stem. In this case, the spiral grain, which can normally prevent living trees from delamination (rope effect) has forced the tree's self-destruction after cutting. Since the tree shapes its design in accordance with the requirements of the living tree during its lifetime, this cracking after felling, of course, does not contradict the principles of self-optimization.

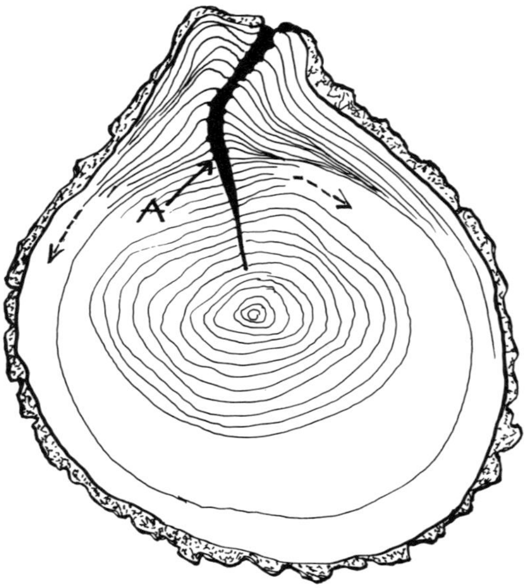

FROST CRACKS AND FROST RIBS - ENDLESS REPAIR AND DESTRUCTION

Oak trees, but other trees as well, are often found with axially bulging ribs some metres in length. Such ribs are the manifestation of a repeated, unsuccessful healing of frost cracks. The ribs' bulging thickness increases with each fruitless effort, and gives an idea of the length of a tiresome chain of frost crack repairs in spring and summer, and destruction of repaired wounds in winter. The open winter crack shown in the drawing seems to contradict all previous explanations of compressive circumferential growth stresses and their closure effects at the periphery (near the bark) on green-stem radial saw cuts. However, with the above-said remaining true, one finds that very low temperatures may convert circumferential compressive stresses into tensile stresses. The causes of such frost stresses are too complex to be outlined here. A detailed description is given by Kübler (1983). Frost crack healing is a very time-consuming process which spoils the fine shape of a tree wherever ribs start bulging. Any first frost crack (A) will only open up at very low temperatures, and be closed again by growth stresses as soon as temperatures rise in spring. Even in its closed state, this crack induces compressive notch stresses at the outermost tree ring, which reduces it by a local increase in thickness. A good compressive contact will lead to welding and final formation of a continuous tree ring covering the crack mouth. Provided the year is wet, the ring is thick enough, and the next winter is mild, the bridging tree ring will not burst again, and the wound will remain closed. In case these conditions are not fulfilled, another frost crack will split the outermost tree ring the next winter, and wound healing will start anew. Without success, the tree shown in the drawing has been toiling at this repair for more than a decade.

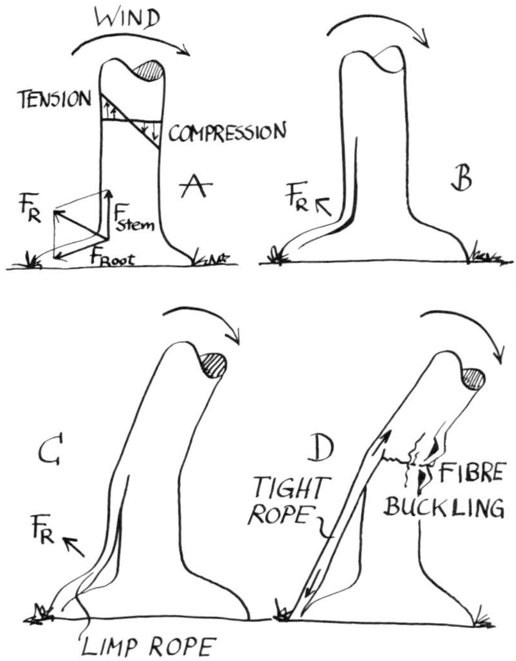

WIND BREAKAGE INITIATED BY ROOT DELAMINATION - A BIOMECHANICAL TRAGEDY

Trees are ingenious, they self-optimize, they master their lifelong task of maintaining a constant stress state on their surfaces, and yet they have one weak point which is highly sensitive and seems to simply trigger failure. This weak point is their insufficient strength to resist lateral tension, and the quasi-inevitable consequence of delamination along the tree rings. Computer simulations by Mattheck and Prinz (1991) have shown that even trees which are well-optimized in the sense of the constant stress criteria have localized peaks of lateral stresses inside, i.e. near their stem-root joints. In a simplified way, the parallelogram of forces in (A) shows that lateral tension has its origin in F_R as a resultant force made up of stem and root forces. The tragedy and treachery about this maximum of lateral stresses is that it is below the surface inside the stem. The "cambium intelligence" in the outermost layer of the tree cannot detect dangers in the inner structure. The tree stops adapting, once stresses are constant on its surface. For lack of sensors inside, it passively submits to heavy wind loads inducing high lateral tensile stresses, which tear the fibres apart and cause delamination (B). The fibres meet again and are held together where the crack ends. In the unloaded state, this separated fibre bundle resembles a limp rope (C). A strong wind blowing will stretch this rope, and the fibres will buckle under abrupt lateral loading at the compression side of the bending, level with the upper end of the rope. Complete rupture finally follows at the tension side of the stem (left side in D). The heavy winter storms of 1989/90 brought about many deplorable biomechanical tragedies. The author saw numerous perfect (in the sense of the constant stress hypothesis), yet sadly defenceless trees at different failure stages and with widely varying crack widths. The sequence of photos published by Mattheck and Bethge (1990) should speak for itself.

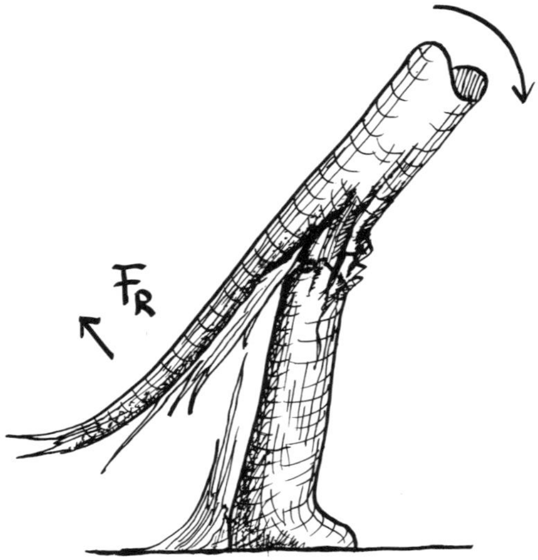

COMPLETE SEPARATION OF UPPER ROOT PARTS AT THE WINDWARD SIDE
(Dr. Mike Coutts, Roslin, stimulated all these considerations on wind breakage)

Failure of the kind described previously can also occur without warning if the entire part of the root is separated by the lateral force F_R. At the moment of complete root rupture, the tree loses much of the mechanical support it gets from the tensile side of bending. Its typical height-of-head failure level with the tip of the crack is assumed to be due to the abrupt reduction of its load-bearing cross-section on the root-rupture side, and the localized wood fibre buckling induced by this loss on the same level at the compression side of bending. The reader should note that this sudden failure is somewhat different from the stepwise process described in the previous example. In the case of sudden failure, the tensile part of the tree (left side) is ruptured all at once. The stem breaks at the tip of the axial crack. The cross-section is thicker above the crack than below it. In the case of stepwise failure, just the opposite is true. The tensile member (rope) on the left of the tree in the previous example reinforces the delaminated part. The delaminated butt below is stiffer than the cross-section above the dividing crack. Speculating on further mechanisms behind the typical height of rupture of trees, one finds that the straight run of fibres is often disturbed by dead branch stubs preferring this height. The bend of fibres around such stubs increases their susceptibility to buckling, and accelerates failure of the in any event weakened stem. Computer simulations at the Karlsruhe Nuclear Research Centre are expected to yield further details to explain this failure mechanism.

WHEN DOES COMPLETE ROOT DELAMINATION OCCUR?

The drawing on the left shows a crack which was initiated by lateral stresses. It runs both up-stem and in the direction of the tip of the horizontal root, but is caught and stopped by a vertical sinker root before it can reach the tip. Under wind loads, the diverted crack tip is compressed, and the crack is arrested. In the drawing on the right, appropriate sinker roots were either absent, or the crack was located near the upper side of the root and was not caught by the sinker root. It is easy to see that the tree on the right will fail through instantaneous root delamination, while the one on the left will be damaged in the course of a slower, stepwise process of the kind described previously.

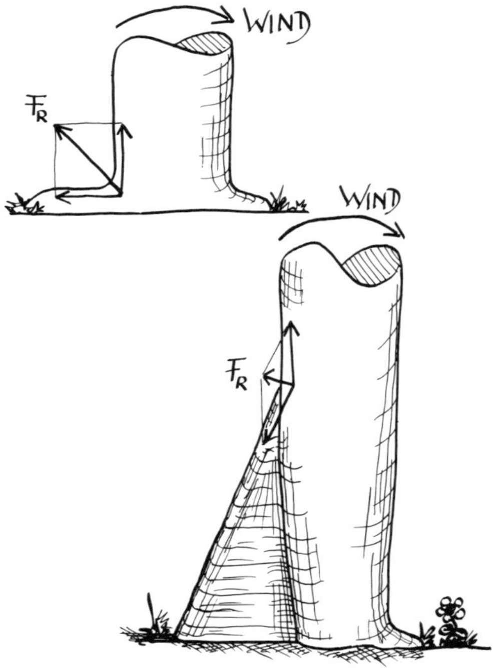

BUTTRESS ROOTS - RELIABLE SHIELDS AGAINST ROOT DELAMINATION

Buttress roots are typical of flat-rooting trees without deeply anchored tap roots. Their main advantage is their ability to stabilize the tree by connecting the outermost lateral ends of the roots with the stem. An advantage, in addition to this lever-arm function, is revealed by the two drawings on the left. Although both trees are under similar tensile stresses due to wind bending, they behave differently. Regarding the acute angle of the root-stem parallelogram characterizing the tree without buttress roots, and the obtuse angle formed by the upper edge of the root and the stem in the tree with buttress roots, it becomes evident that the latter is protected from root delamination due to this obtuse angle, which reduces the force F_R. With large protective buttress roots on their windward sides, none of the black poplar logs examined by the author was found to have failed as a result of root delamination. Instead, they all broke due to complete internal decay, which weakened their butts or even the very buttress roots. Stimulation of buttress root formation would be a promising method of reducing the danger caused by wind breakage initiated by root delamination. Among the solutions investigated is the far-stem arrangement of sinker roots outlined in one of the previous examples. Selective breeding would be another effective, yet time-consuming method. Mechanical suppression of centred sinker roots, for example by planting the tree over a wooden disc or by strangulation of the juvenile tap root, sounds feasible but would be expensive. Considering the disastrous effects of wind breakage due to root delamination, it is worthwhile continuing to look for adequate, universally applicable solutions, which avoid major losses.

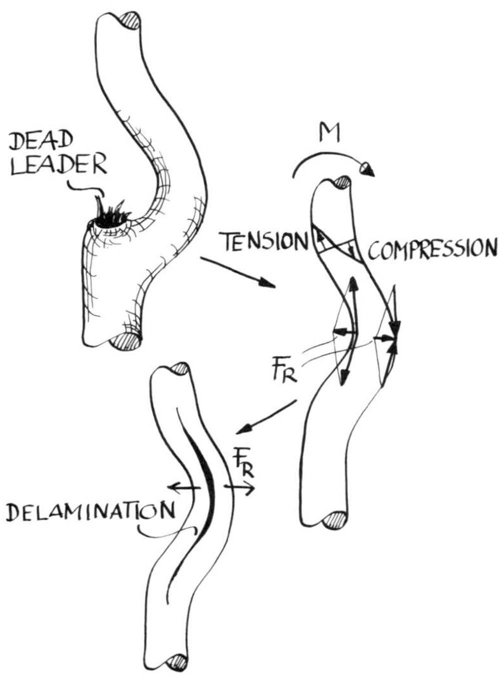

DELAMINATION OF TREE KNEES

Previous examples have shown how a lost leading shoot is replaced by a suitable side branch brought upright by negative geotropic corrections, which replaced the centre of gravity of the crown over the centre of the root stock, and avoided unnecessary tree butt bending. With regard to what happens if tree knees formed by side branch erection are subjected to wind bending, one finds that straightening of a knee is accompanied by lateral outward forces, which because of the insufficient lateral strength of wood can cause axial cracking in the centre of the knee. Finite element calculations performed by Mattheck and Burkhardt (1990) at the Karlsruhe Nuclear Research Centre have shown that as soon as axial delamination sets in, shape optimization of the outer tree design also completely fails. This means that the constant stress state, which the tree has adopted and taken care of during its lifetime by adaptive growth, is ruined in just one moment. Notch stresses are therefore high in the concave areas of the outer contour after axial cracking, and present an additional failure risk. Since delamination of that kind may well be found with any bent wood fibre bundle exposed to tension or bending, tree knees and bent beam sections involve a serious, high potential risk to trees. These were termed *hazard beams* to express the threat they represent to any tree.

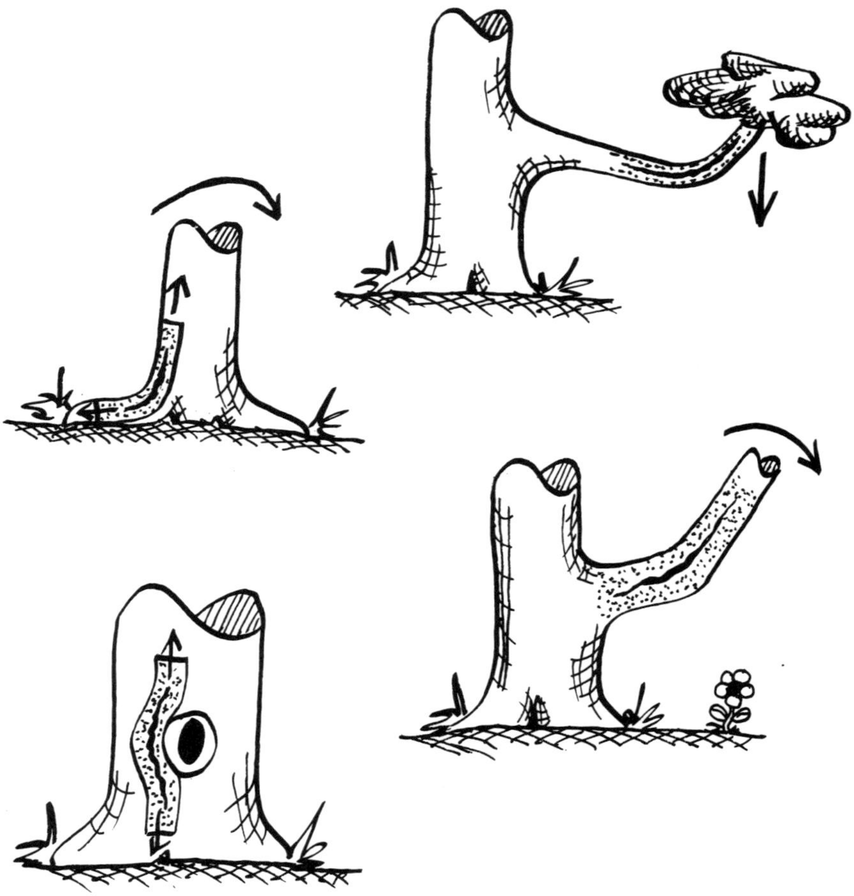

HAZARD BEAMS: THEIR LOCATION

The failure case described previously can start from many different points inside the tree. It seems that healthy trees are nearly exclusively damaged by mechanisms which are at least initiated by axial delamination of bent wood fibre bundles, namely *hazard beams*. Damage is caused by lateral stresses which only act inside the fibre bundle and, being zero on the surface of the tree, cannot be detected by the cambium. Computer calculations made at the Karlsruhe Nuclear Research Centre have shown that adaptive growth does not reduce these local lateral stress maxima. In a self-explanatory way the examples (dotted areas) illustrate the variety of forms and manifestations of these damaging, fatal *hazard beams*.

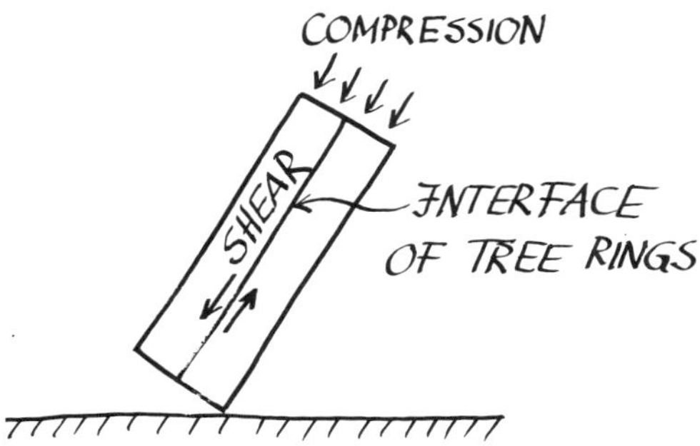

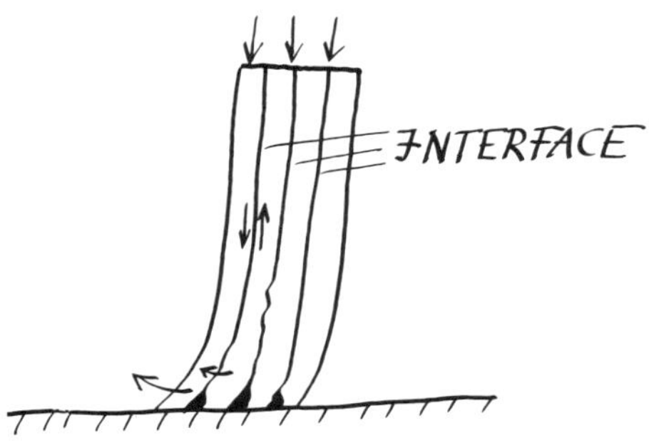

SHEAR STRESSES BETWEEN TREE RINGS - A RISK AND ITS NATURAL AVOIDANCE

Annual tree rings consist of mostly somewhat softer and more porous early wood (spring wood), and stiffer and denser late wood (autumn wood). One can easily imagine that with their sensitive early wood in between, tree rings located side by side are likely to delaminate or be torn along their interfaces if exposed to shear stresses. The two examples on the left show high shear stresses caused by pressure, which is non-perpendicular relative to the contact surface. In response to such critical situations, tree rings align along the shear-free principal stress trajectories, that is, the force flow. This axicircumferential ring--trajectory coincidence avoids the risk of shear stresses, and bending and shear are converted into tension or compression. It is worth mentioning that trabecular bone arranges its microframeworks according to the same trajectory design rules. The following examples will show that newly formed tree rings even adapt to diverted force flows. Older tree rings formed before loads were changing will not adapt. Bone is of higher quality in this respect because it also changes older structures inside. Its higher adaptiveness, however, conceals the interesting mechanical load history that tree rings are able to reveal. An excellent book on tree rings by Schweingruber (1989) presents a variety of impressive examples.

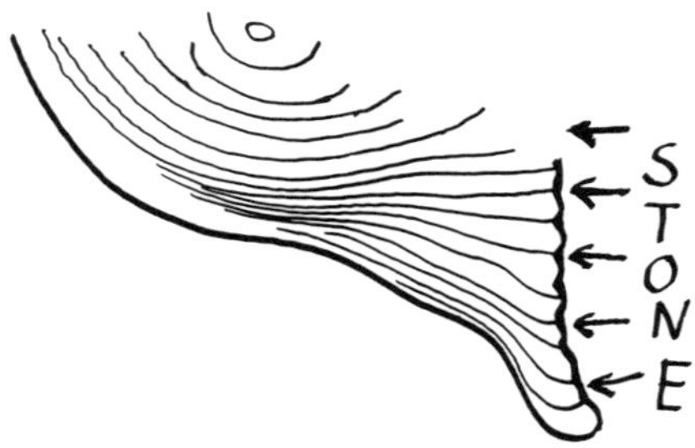

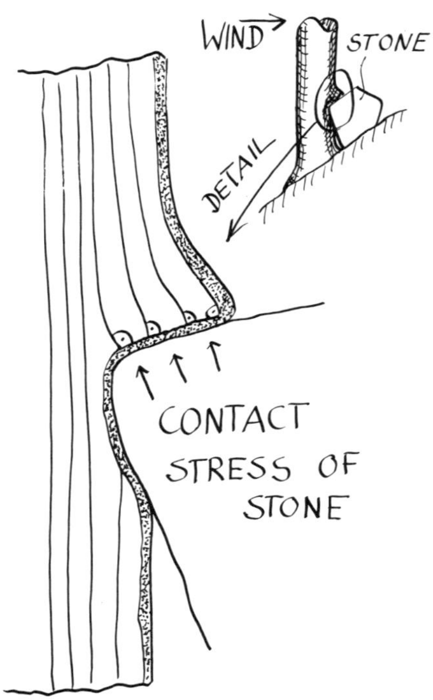

TREE RINGS ADAPTING TO CONTACT LOADS CAUSED BY A STONE

As described already, the constant stress state on the surface of a tree is disturbed if a stone rolling downhill is stopped by the tree and contacts it. Contact stresses are reduced as the tree grows around the stone and enlarges the contact surface. The parallel-to-the-grain cut illustrated on the left shows that tree rings are thicker in the contact zone, and tend to run perpendicular to it. Computer simulations performed by Mattheck, et al. (1989 a) at the Karlsruhe Nuclear Research Centre have been showing excellent agreement of force flow and axial tree ring orientation. This coincidence avoids the dangerous shear stresses between the tree rings and bending stresses within the tree ring, and aligns the wood fibres to the force flow. The circumferential tree ring formation revealed by the perpendicular-to-the-grain cut shown above has been furnishing proof that tree rings also follow the path of the circumferential force flow.

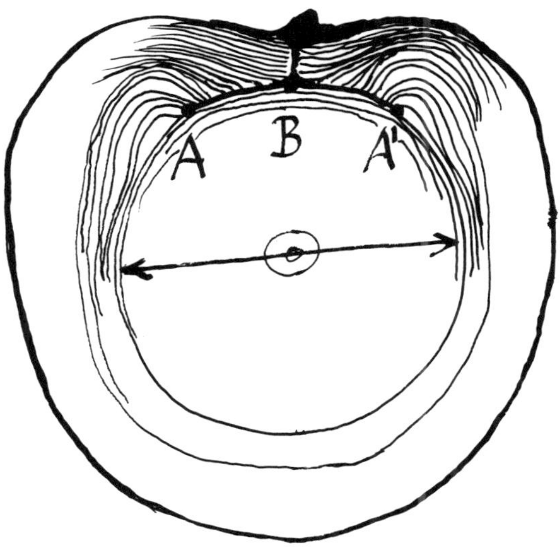

SAW CUT THROUGH THE HEALED WOUND OF A PINE

Any wound is also a notch from the mechanical point of view. The force flow has to be diverted around this notch, and localized stress peaks (notch stresses) have to be expected, which disturb the state of constant stress on the tree surface. The tree starts enlarging its developing, latest tree ring in the area of the notch edge to restore the constant stress state and repair the weak point in its design. The wound extends from A to A'. The arrow indicates the diameter of the tree at the time of wounding. The tree ring width beside the wound is several times the width of the rings farther away from it. In the initial wound healing phase, the shear stresses acting between the developing tree rings are minimized by tree ring arrangement perpendicular to the wound surface. As healing progresses, the wound callus coming from A and A' grows towards the axis of symmetry (above B) and meets there. This is when tree rings arrange themselves perpendicularly to the bark squeezed between the callus (dark vertical line above B). In a few years the outermost tree rings coming from both sides will "melt" the bark and weld, and the wound will be closed by a continuous encircling tree ring.

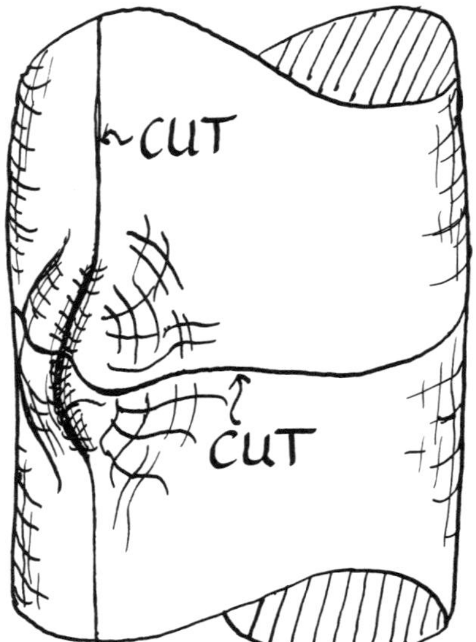

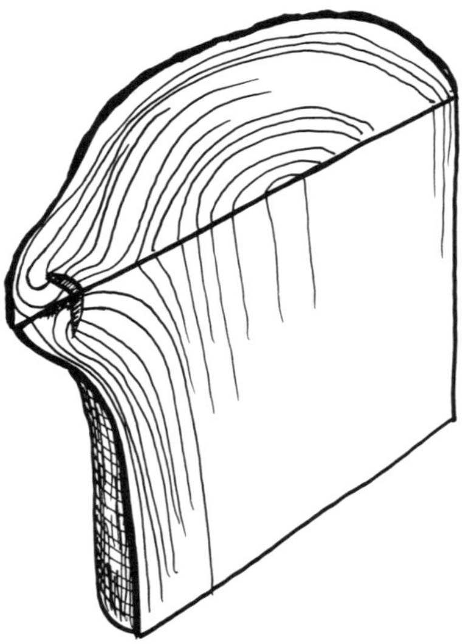

A COMPLETELY HEALED WOUND

The two continuous, latest tree rings running around the notch in this quartered log show that wound healing was already completed 2 years ago. Unlike the rings in the former pine wound example, they do not end at the enclosed bark, but cover the former wound without interruption, and prevent decay. The somewhat bulgy shape reveals that this rapidly closed wound stems from an old or weak branch, which was surrounded by a collar forming a ring notch. The branch broke or was cut off at this predetermined breaking point. The flat and more concave wound shown in the previous example does not appear to be such a branch hole.

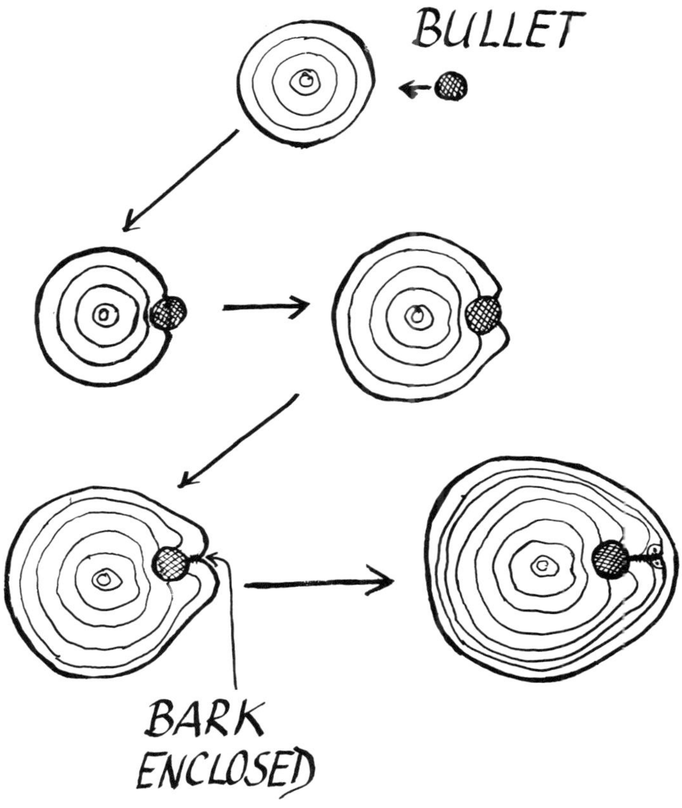

BULLET ENCLOSURE - A BIOLOGICAL DIARY OF HISTORY

In sawmills, one often sees old stems with variously sized metallic inclusions. Although most of these particles are heavily attacked by the corrosive saps of the trees, the position they have in the tree ring structure reveals interesting historical facts, which are confirmed by the workers who have identified them as fragments of shells or grenades. The sketch shows the successive integration of a low-energy shotgun pellet which did not penetrate deeply, but became more and more enclosed by tree rings, increasing their thickness near the wound. The tree rings approach the bullet from both sides, and soon cover it up by fusing. Wound healing is completed, and an increasing number of tree rings will cover and hide the wound and the bullet. Deeply penetrating high-energy projectiles perforate existing tree rings. According to the shapes and sizes of such bullets, tree rings formed later may vary in shape, but the principal process of bullet integration is similar as long as the bullet is in contact with the cambium.

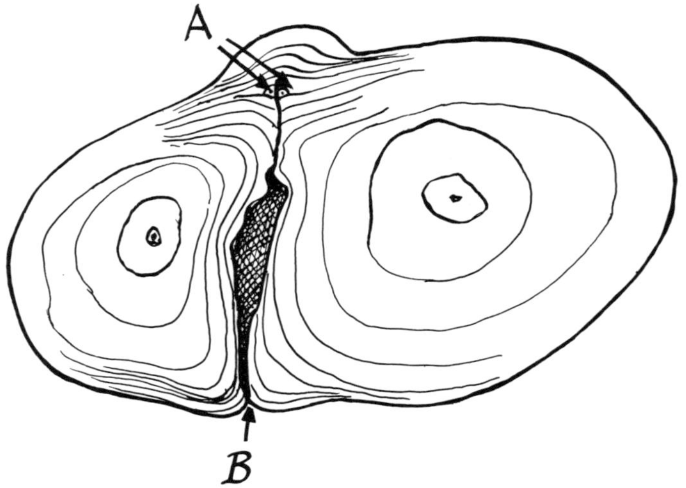

SAW CUT THROUGH A COMPRESSION FORK

Compression forks were already introduced as being twin stems, which due to insufficient light reception in dense stands are pressed against each other by their own reaction wood. The two stems form a nearly plane, enlarged contact surface to reduce the contact pressure between them. They increase the widths of their tree rings at the edges of the interface, and progressively arrange perpendicularly to the bark dividing the interface. The tree rings align with the force flow, and by and by break the bark as they develop continuous circumferential contours. At (A) the tree rings are already perpendicular to the interface and have fused. At (B) it will take no more than another year for them to weld. Although fusion is completed at (A), contact stresses seem to persist due to further formation of reaction wood. The fork increasingly narrows to bring the stems upright and ever closer to the light. This leads to the development of a bulgy axial rib or ridge, which is rather typical of the compression forks of both gymnosperms and, perhaps even more impressively, angiosperms. Especially beeches may form ribs as thick as the stem itself. Compression forks are twins hindering each other. They are an example of the keen competition between neighbours and close relatives in dense stands, which are too dark for peaceful coexistence.

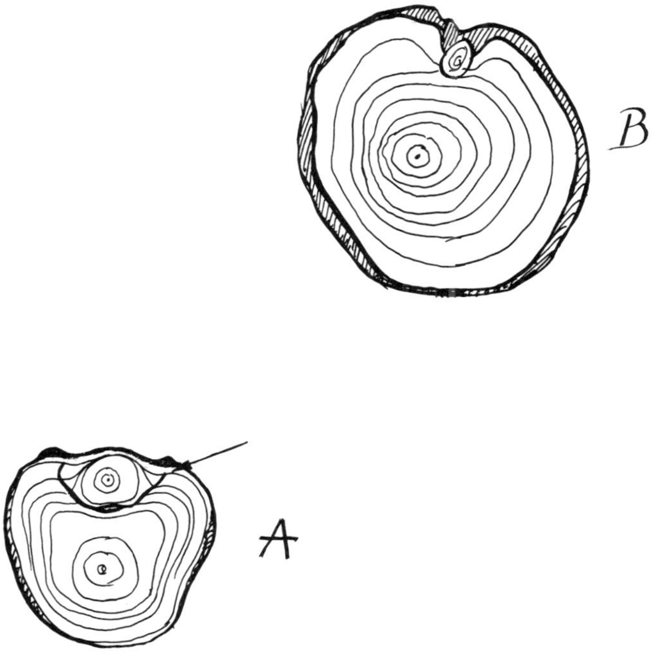

TREE RINGS AROUND DEAD AND LIVING BRANCHES

Both sketches, (A) and (B), illustrate the integration of a branch, which protrudes forming an acute angle with the main stem. Although similar at first sight, integration proceeds in two totally different ways. The branch (A) is still alive. It is able to attach a new tree ring to join the one produced by the stem into the acute angle. As soon as the two rings contact, the contact surface is enlarged in the usual way by a local increase in tree ring thicknesses. The branch starts to ovalize, and the stem grows around it from both sides. They fuse as soon as their tree rings meet on the far-stem side of the branch in a non-kinked, straight way (arrow). A common, continuous force flow starts running through the tree rings enclosing both stem and branch. In the course of time, the living branch will be increasingly enclosed from the base upwards. Unlike the living branch, the dead one (B) cannot fuse with the stem because its cambium is no longer active. It is treated like the bullet described previously, that is, it will be enclosed like a dead or foreign body.

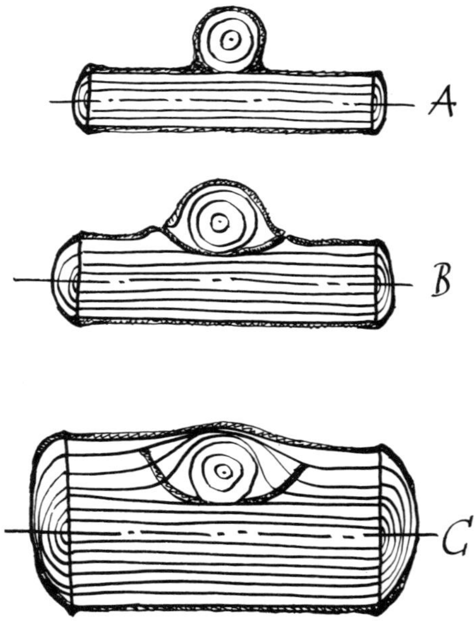

TREE RING ARRANGEMENT IN CROSSED WELDING PARTNERS

Two cospecific branches of similar thicknesses contact crosswise (A). Their contact pressure, combined with wear due to wind movement, is high enough to stimulate enlargement of the contact zone. They start growing around each other, and join in a well-matched form fit (B). After a period of 3 years this form fit is so perfect and stable that relative movements are minimized. The upper branch has grown into a pointed oval (C), and its latest tree ring is in axial alignment with the latest upper tree ring of the branch below it. The branches unite to melt the bark in a joint effort and form a common tree ring. Because kinks within the now fused tree ring are totally avoided this way, the tree ring is loaded predominantly by tension or compression but no longer by bending. Also, the shear between the individual tree rings is minimized by their orientation along the force flow. Ideal as it may be, their intimate fusion is marred by a handicap imposed upon the upper branch. In most cases, it is partially cut off from the transport system and has to be nursed below the welding by its fused partner. The natural equivalent of this constructed case is described in the following example.

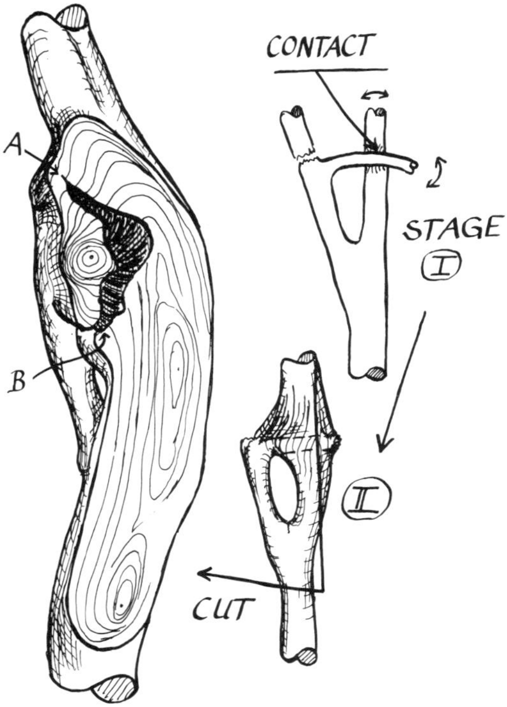

TREE RING ARRANGEMENT IN A WELDED FORK

A branch originating from the left part of a fork contacts the right fork member (I). Frictional wear causes a wound, and the contact stresses induced are minimized by contact surface enlargement. Wound healing and form fit processes stabilize the new joint so that movement of the contacting partners relative to each other is soon controlled. Welding is completed by a vertical common tree ring, which fuses the contacting branches (II). A cut through the fused structure reveals all of the stages required to prepare a perfect fusion. The once circular, adopted branch starts to ovalize and arrange its tree rings perpendicularly to the bark-enclosing interface (shown as a dark, gaping hole in the log-drying phase). However, this perpendicular arrangement of tree rings does not seem to suffice for an early fusion. The branch has to continue producing tree rings in order to align with the dominant vertical force flow. A pointed oval is formed as further rings are added in the course of time. It takes several years for this oval to form, coincide with the force flow, and provide a continuous common tree ring "melting" the bark at points (A) and (B). Branch and fork fuse into one; their common cambium is the biological completion of their mechanical union. The upper part of the left member broke off, and, as is typical of such welded forks, the right part of the integrated branch was cut off from the transport system and died. Smoothed by adaptive growth, the sharp edges formed by the contacting branches are free or at least nearly free of notch stresses in their final phase. The wind-exposed right part of the fork grows thicker above the bridging branch than below, where both parts of the fork are relieved due to the structure's frame effect. The last circular tree ring in the horizontal branch indicates the year of beginning contact. The mechanical loads are evenly distributed in this shape-optimized structure. Its design is again determined by the constant stress criteria, the presumably most general design rule for biological load carriers.

7. CONCLUSIONS

Modelling tree design was shown to be a complex, varied process owing its unique success to a few clear rules. The mechanical laws by which it is governed were introduced presenting a variety of examples provided by nature. The following statements are meant as a brief summary of these major rules and their consequences.

1. Trees are biological load carriers, which self-optimize by adaptive growth.

2. Their optimum mechanical state is reached as soon as mechanical stresses are equally distributed on the tree surface, no area carrying more than another. A weighted time average is assumed for different loadings at different times.

3. If the optimum state of constant stress is disturbed, it will be restored by the tree through the attachment of more material at points of maximum stresses.

4. In this sense, wound healing is most active at points where wounds (notches) induce the highest local stress peaks (notch stresses). Those points which are weakest and are most likely to fail are repaired the most rapidly.

5. In accordance with the above, contact surfaces are enlarged fastest, where contacting foreign bodies cause the highest stresses on the tree surface. Contacting bodies are enclosed, the tree grows around them, or even integrates them. Dead branches of the same tree or parts of different tree species are treated as dead or foreign bodies, e.g. stones.

6. Contacting, living parts of trees of the same species may weld and form frames or framework designs, which support the tree, and are characterized by their efficient use of material. Normally, the formation of common tree rings takes place firstly if the tree ring runs perpendicularly from both sides to the contact surface (no kinks in the alignment), and secondly if the continuous, encompassing tree ring is aligned in the direction of the main force flow.

7. Tree rings align to coincide with the dominant force flow (principal stress trajectories) both axially and circumferentially. This leads to minimized shear and bending and maximized tension or compression loading along the tree rings. Older tree rings cannot adapt to changed load conditions, but reveal the mechanical load history of the tree by providing the observer with information about bygone force flow directions.

8. Besides these efforts to master externally applied loads through evenly distributed stresses, the tree tries to minimize just these external loads by placing its centre of gravity over the centre of the root stock. This is done by forming reaction wood, i.e tension wood in angiosperms and compression wood in gymnosperms. In this way, bending moments due to gravity, and torsional moments due to wind loads, are kept as small as possible. This rule is ignored whenever the tree follows phototropic stimulation to reach more light.

9. The tree reacts to any change in its loading or stress distribution by an adaptive growth response, which modifies its shape. *This modified shape documents the tree's load history - it is its autobiography, its mechanical diary which observers are invited to read.*

To recognize the changes in the design of a tree brought about by the changes in the loads to which it is exposed, is to understand the body language of trees, the language telling their fates.

For the insider who understands this body language, an otherwise monotonous, lifeless hornbeam forest starts whispering stories of injuries and healed wounds, of contacting branches growing around each other, of weldings, unions, rivalries, of light and darkness, of friendship, strength, or weaknesses. An ugly branch hole in the door of one's room will suddenly turn into a mysterious messenger unveiling its secret of force flows, shear stresses, mechanical impacts - revealing even its year of death.

The world which is disclosed to us in this way is a silent world. It is the soft monologue of a mechanical body language, which can further our understanding of the difficult struggle for survival that the individual organism must lead against the merciless trials of nature.

REFERENCES

Kübler H (1959 a) Die Ursache der Wachstumsspannungen und die Spannungen quer zur Faserrichtung. Holz als Roh- und Werkstoff 17:1-9

Kübler H (1959 b) Studien über Wachstumsspannungen des Holzes, Zweite Mitteilung: Die Spannungen in Faserrichtung. Holz als Roh- und Werkstoff 17:44-54

Kübler H (1983) Mechanism of frost crack formation in trees - a review and synthesis. For Sci 29:559-568

Kübler H (1987) Growth stresses in trees and related wood properties. For Abstr 48:131-189

Kübler H (in preparation) Function of spiral grain in trees

Mattheck C (1988) Warum sie wachsen, wie sie wachsen - die Mechanik der Bäume. Rep No. 4486, Karlsruhe Nuclear Research Centre

Mattheck C (1990 a) Engineering components grow like trees. Materialwiss Werkstofftech 21:143-168

Mattheck C (1990 b) Why they grow, how they grow - the mechanics of trees. Arboricult J 14:1-17

Mattheck C, Bethge K (1990) Wind breakage of trees initiated by root delamination. Trees - structure and function 4:225-227

Mattheck C, Burkhardt S (1990) A new method of structural shape optimization based on biological growth. Int J Fatigue 12:185-190

Mattheck C, Burkhardt S (1991) Der Unglücksbalken. Intern Rep, Karlsruhe Nuclear Research Centre

Mattheck C, Korseska G (1989) Wound healing in a plane (Platanus acerifolia, (AIT.willd.)) - an experimental proof of its mechanical stimulation. Arboricult J 13:211-218

Mattheck C, Prinz M (1991) Buttress roots: why they grow and what they are good for. Intern Rep, Karlsruhe Nuclear Research Centre

Mattheck C, Vorberg U (1990) The biomechanics of tree fork design. Intern Rep, Karlsruhe Nuclear Research Centre

Mattheck C, Walther F (1991) Eine neue Fälltechnik zur Vermeidung von Hirnrissen in Laubbäumen. Intern Rep, Karlsruhe Nuclear Research Centre

Mattheck C, Huber-Betzer H, Keilen K (1989 a) Die Anpassung eines Baumes an die Kontaktbelastung durch einen Stein. Rep No. 4526, Karlsruhe Nuclear Research Centre

Mattheck C, Walther F, Keilen K (1989 b) Hirnrisse in Laubbäumen - ihre Entstehung und Wege zu ihrer Vermeidung. Rep No. 4557, Karlsruhe Nuclear Research Centre

Mattheck C, Walther F, Keilen K (1989 c) Ein Kompressionsring zur Begrenzung von Hirnrissen in Laubbäumen. Rep No. 4629, Karlsruhe Nuclear Research Centre

Mattheck C, Huber-Betzer H, Keilen K (1990) Die Kerbspannungen am Astloch als Stimulanz der Wundheilung bei Bäumen. Allg Forst Jagdztg 3:47-53

McMahon Th (1973) Size and shape. Science 179:1201-1204

Metzger K (1893) Der Wind als maßgeblicher Faktor für das Wachstum der Bäume. Mündener Forstliche Hefte, 3. Heft, Berlin

Schwabe A, Kratochwil A (1987) Weidbuchen im Schwarzwald und ihre Entstehung durch Verbiß des Wälderviehs. Ungeheuer und Ulmer, Ludwigsburg

Schweingruber FH (1989) Tree rings - basics and applications of dendrochronology. Kluwer, Dordrecht

Volks T (1990) Computersimulation der Wundheilung bei Waldbäumen. Intern Rep, Karlsruhe Nuclear Research Centre

Walther F (1990) Computersimulation adaptiven Wachstums biologischer Strukturen: Gestörter Baumwuchs. Intern Rep, Karlsruhe Nuclear Research Centre

Wilson BF (1984) The growing tree. Univ Press, Amherst, Massachusetts

Zimmermann MH, Brown CL (1980) Trees - structure and function. Springer, Berlin Heidelberg New York

Zimmermann MH, Wardrop AB, Tomlinson PB (1968) Tension wood in the aerial roots of Ficus benjamina L. Wood Sci Tech 2:95-104

SUBJECT INDEX

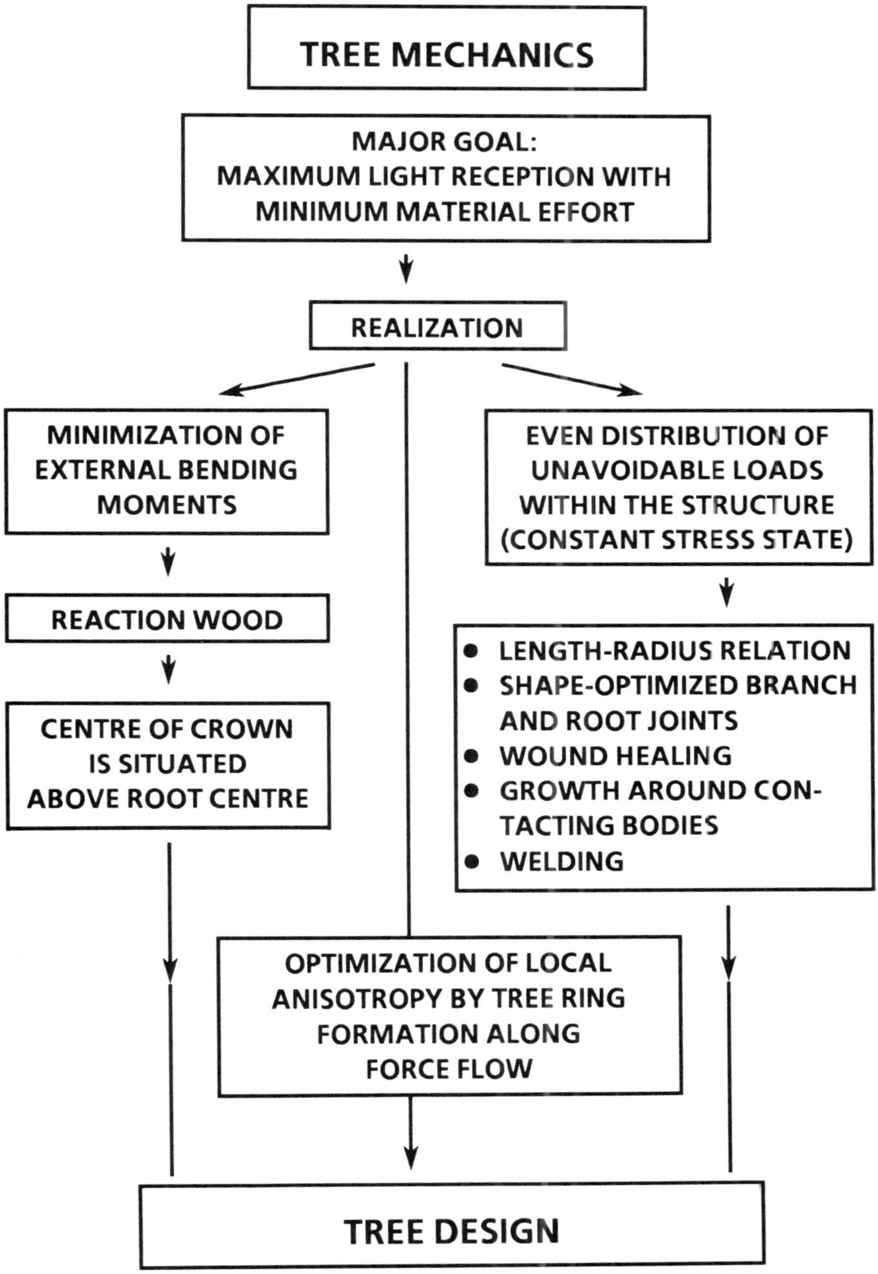
TREE MECHANICS
MAJOR GOAL: MAXIMUM LIGHT RECEPTION WITH MINIMUM MATERIAL EFFORT
REALIZATION
MINIMIZATION OF EXTERNAL BENDING MOMENTS
REACTION WOOD
CENTRE OF CROWN IS SITUATED ABOVE ROOT CENTRE
EVEN DISTRIBUTION OF UNAVOIDABLE LOADS WITHIN THE STRUCTURE (CONSTANT STRESS STATE)
LENGTH-RADIUS RELATION
SHAPE-OPTIMIZED BRANCH AND ROOT JOINTS
WOUND HEALING
GROWTH AROUND CONTACTING BODIES
WELDING
OPTIMIZATION OF LOCAL ANISOTROPY BY TREE RING FORMATION ALONG FORCE FLOW
TREE DESIGN